THE
GREAT
SUNDIAL
CUTOUT
BOOK

THE GREAT SUNDIAL CUTOUT BOOK

Robert Adzema

Mablen Jones

Photographs of the models by
Rosmarie Hausherr
Illustrations by the authors

HAWTHORN BOOKS, INC.
Publishers / New York
A Howard & Wyndham Company

THE GREAT SUNDIAL CUTOUT BOOK

Library of Congress Catalog Card Number: 78–52964
ISBN: 0–8015–3117–9
1 2 3 4 5 6 7 8 9 10

Contents

THE GREAT SUNDIAL CUTOUT BOOK

Shêng, **sundial sculpture by Robert Adzema.**

Introduction:
Our Lust for Light

People often ask Robert Adzema and me how, as artists, we became interested in sundials. This query invariably produces a flood of associations, a chain of memories of influences and ideas. We reflect on our personal artistic development, on what nourished and sustained us. However, there are always two answers (one for each of us) that stand out from the rest.

Bob came to sundials as a direct result of his interest in developing a public outdoor art. He wanted to know and understand how a sculpture in a plaza or park would react to the continually changing sunlight from day to day, from season to season, and from place to place. He studied the influence of shadows on the way forms look in different latitudes—for example, how deep the undercutting of shapes would have to be for a sculpture in Boston as opposed to one for Phoenix, Arizona, in order to be read from the same distance (for the sun's elevation in the sky is higher for places closer to the equator).

The path was less direct for me, and came through related themes of both my sculptures and paintings. While working on a series of terra-cotta sculptures, I researched how physical forces mold the landscape. I discovered that pullings of astronomical bodies and electromagnetic flows of particles from the sun contributed to shaping our terrestrial forms. My next series based on

Geological landscape sculpture in terra-cotta by Mablen Jones entitled *Mid-Atlantic Ridge.*

meteorology again brought me back to the sun and its influence on our weather.

Although the space adventures and moon walk of the late 1960s inspired our interest in astronomy, it took an increased awareness of our personal place on earth to turn our focus toward the sun and sundials. In the early seventies a nearby university asked Bob to submit a sculpture proposal for their campus plaza. He decided to make a unique piece for that particular spot, not one that could merely be set down anywhere as decoration. Bob wanted the work visually to express and dramatize the site's physical coordinates in time and space. Although the original renovation plans for this tree-lined mall called for a clock tower, Bob felt that the timepiece for this place should be more cosmic and came up with a design for his first sundial sculpture, entitled *The Tower of the Sun*, which resembles a giant futuristic space flower from another galaxy.

The lack of continuous natural sunlight in our old studio was one of the most exasperating irritations Bob overcame while developing this piece. That New York City garage, which extended under and beyond the back of a three-story building, had only two tiny windows facing the smoke vent of a short-order Portuguese restaurant next door. Finally, after working on that dial for several months, on the most frigid gray day in December when the daylight was nearing the shortest limit of the year, Bob, in total desperation, put up a ladder, cut a hole in the roof to get at the light, and installed a skylight. As he finished cutting and pushed through to the open sky, the snow poured in. Our life with limited daylight seen through perpetually smoke-greased windows had made us a bit unaware of the daily meteorological conditions.

That move still didn't satisfy his lust for light on the dials, for surrounding buildings towered above ours and blocked direct sunlight for many hours of the day.

So Bob next invented the Sun Simulator, a machine carrying a light on a toy car that crawled up one side of the loft, then across the ceiling and down the other side on a movable arc. The parts were adapted from a plastic Japanese roller-coaster-type toy called the Grippedy Gravity. Bob's brother Russell, an experienced electrician, supplied the lighting expertise to run the sun. A crossbar was attached to the ceiling so that the track could be tilted to simulate the altitude of the sun for each season of the year. Now the models could be pretested with simulated sunlight for any time, both day and night.

Unfortunately, after Bob submitted the model, the college fathers decided that they lacked sufficient available money to develop the square properly, so no artwork has been placed there to this day. Nonetheless, the making of that one sundial sculpture introduced us both to the mysteries and motions of the sun. Bob discovered that the geometric forms arising from the dial's functional use were intrinsically beautiful themselves, and he has since been absorbed by the adventure of joining together sculpture and dial—a vital theme that he has found inexhaustible.

As I continued to research more about the astronomy pertaining to the subjects of my own art, I uncovered a wealth of images,

Sundial sculpture by Robert Adzema entitled *The Tower of the Sun*. Photo by Rosmarie Hausherr.

ideas, information, and myths about the sun. The idea for Bob and me to combine our common solar interests and knowledge into some kind of collaborative project seemed natural.

Why a cutout book? Because our involvement with miniature paper models and sketches has always been an inseparable part of our creative processes in both sculpture and painting. We cannot afford to experiment with large, expensive, permanent materials to build full-sized works without preparation. Countless paper maquettes and drawings accompany the development of any of our ideas for pieces. When one solution doesn't come out exactly right, we fold and cut, redraw, and tape or glue on. We may inadvertently end up with collage, origami (Japanese paper folding), or découpage (decorative paper cuttings) by default.

Moreover, we have found that small is just as beautiful, sometimes even more so, than large. The miniature scale has a magic

Studio showing the beginnings of the Sun Simulator, sundial sculpture, and desperation skylight.

of its own that lends itself to visions of the universe. You can create a portable cosmos; twirl your celestial world (see the armillary sphere in chapter 5) on a string. There's something about the miniature that evokes the immense. A world within a tabletop terrarium conjures up a jungle; a small aquarium calls forth an ocean. Real-size objects become prosaic, have the feeling of the expected, are the furniture of practicality and mundane routine. Here in our miniature microcosm we have a panoramic view, and we get an Alice in Wonderland sensation that we've just eaten the mushroom and become ten feet tall, a reverse of that childhood intimidation when everything was too high above us and the world inhabited by giants.

These experiences also demonstrated to us that you don't have to cut holes in your roof to enjoy sundials. The miniature paper dials we constructed were radiant during the limited morning periods when direct sunshine penetrated our two tiny windows, and the models were portable enough to carry to the park or beach. While you may need many hours of sun to work on inventing and testing the dials, you require much less of it merely to enjoy them. The effect of certain hours is more dramatic on some models than on others.

Bob and I found to our surprise that many people see only the utilitarian function of sundials. The first things they usually ask are "Does it work? How accurate is it?" They rate its value without knowing what the dial really does and why it doesn't tell clock time. We hope that in making these models you will come to recognize their other virtues and uses as well—their grace as vehicles for flights of the imagination, their elegance of form for sculpture, their illumination of celestial phenomena and the sun, and of course—their fun.

Mablen Jones

5

The Sun Bather by Robert Adzema.

I
Sun Lore of Many Ages

Centuries ago, time, society, and the cosmos were inextricably linked. Now we live in a schism between intuited biological rhythms and chronarchy (rule by the clock). Our past is broken into discrete events where only beginnings and endings recall an order, such as births, deaths, graduations, and other rites of passage.

The digital watch offers us a dynamic image of our fractured contemporary vision of time. We view the numbers much as we observe the events of our days—like digits flashing, then disappearing into oblivion and existing only as we recall their passage. At least the traditional clockface, with hands orbiting about the dial in a circular path, gave a map, a direction, and a trajectory for a journey through time, a connection between past and future. Indeed, that circular image dominated much of history's ideas about the gigantic clocklike workings and spatial composition of the universe. The flashing digital timepiece parallels a more contemporary cosmological theory that matter and energy in the universe are oscillating events constantly alternating from one state of being to the other.

However, our physical bodies and all forms of organic life above the level of viruses and bacteria are running on a different clock—that is, solar time. Solar rhythms result from daily, seasonal,

yearly, and longer periods of cyclical movement of both the earth and the sun in relation to each other, and they vary in length, speed, and quality relative to our celestial position in space. The sun is our biological and emotional pacemaker.

The sundials and toys we offer in this book celebrate this great god of life and poetry and light—our star of stars. These dials are brought to life by the sun's radiant aura, the same emanations which bring us comfort, warmth, and healing. We turn toward the light flooding into our rooms, entranced by the lengthening shadows creeping over dial, sill, and stair. Such moments are "time out," an escape from the regimentation of the clock. Estrangement from the solar pace and processes is neither healthy, comfortable, nor ecologically sound—for our moods and even our state of health change with the seasons and the weather.

We all see nature's sun clocks in the obvious annual patterns of plant and animal growth and in such activities as migration and hibernation. In addition, our many unseen daily and seasonal body rhythms are connected with the sun. For example, we feel "highs" in the spring and fall, which coincide with peaks of mental acuity, and "lows" in midsummer and late winter. For several years I dated my artworks and found that creative periods recurred cyclically, particularly starting just after the winter solstice. Sea-

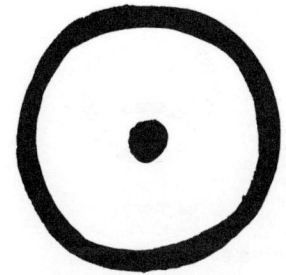

7

Sumerian   **American Indian**   **Egyptian (eye of sun god)**   **Oriental (rising sun)**

Examples of sun symbols

sonal fluctuations of the reproductive cycles of both animals and man are also stimulated by the color and amount of light. The hypothalamus of the brain as well as the pineal gland and endocrine system use the stimulus to regulate fertility cycles. Our body temperature drops and metabolism slows down at night even when we're not sleeping and when the temperature of our surrounding environment stays the same. Our blood coagulates, or clots, faster at the coming of dawn and when sunspots (dark patches on the sun's surface that are strong electromagnetic fields) pass across the center of the sun, and the coagulation rate declines during solar eclipses, when the moon blocks the sun from the earth. The gamma globulin portion of our blood serum, which contains most of the immunizing antibodies that fight infection, drops by as much as 28 percent at night. Other similar sun-affected rhythms are the white blood cell count, the serum iron, and blood sugar levels. Our heart rate, blood pressure, and excretion of urine, phosphate, potassium, and sodium are also solar influenced.

The Invisible Powers of the Sun

The causes of these and other natural phenomena that we will describe later are both visible sunlight and the invisible solar wind. The solar wind results from the sun's corona (or outer atmosphere) expanding and blasting out a stream of atomic particles. It varies in intensity but blows constantly in waves from the sun out into space at a rate of approximately two hundred to five hundred miles, or roughly three hundred to eight hundred

kilometers, per second. We do not consciously feel it, although our blood composition and brain-wave cycles register its effects. Most of this wind is deflected by our earth's magnetic field, although fast particles do enter the atmosphere more directly in the polar regions. This flow is made up of electrons and protons within a substance called a plasma. Plasma is a substance similar to a gas in which some electrons and protons are knocked out of their nuclear orbits and freely move about creating electromagnetic fields. This solar wind also includes invisible ultraviolet light, infrared rays, X rays, and longer waves called corpuscular radiation. The wind velocity is influenced by magnetic differences in the sun's field. Sunspots are areas of extremely strong magnetic intensity, and flares break out in areas of large groups of these spots. These flares send forth intense ultraviolet waves.

Our bodies are responsive receptors of these electromagnetic waves from space, and our nervous-reaction times are slowed down after the great flares from sunspot areas. The wind of the sun also influences the movements of high- and low-pressure weather systems, which depress our bodies with positive ions and stimulate them with negative ones. We feel better and are more efficient on clear, sunny, high-pressure days than on gray, stormy, low-pressure ones.

The irregular waves generated during the sunspot cycles leave other physical evidence engraved in the annual rings of trees, in the number of droughts, earthquakes, icebergs sighted, and in the great plagues and epidemics of history. Heart attacks (due to the raised clotting index) and accidents (due to slowed nervous-reaction times) are more likely just after periods of violent solar activity.

8

The electromagnetic radiation, which takes about two days to reach us from the sun, affects radio reception, chemical reactions, and the freezing point of water (according to the position of the earth within the solar field). In addition, the polar northern lights, or *aurorae boreales* (and southern lights, or *aurorae australes*), are caused by the solar wind hitting and agitating the magnetic fields around the poles.

I recently met an acquaintance who is a financial analyst for a bank and asked her what she was working on. She replied, "You're not going to believe this. We're charting sunspot cycles to try to predict economic trends." I believed her—our relationship to our star may be the most important factor governing our lives in the coming decades. Naturally, we will be using solar energy for daily comforts and necessities such as fuel, solar medicine and psychiatry, and weather control, but an understanding of the sun's movements will be a necessary tool in shaping the architecture of the future and our lives. Instead of merely learning national and international geography in school we shall study interplanetary and intergalactic maps as well. For space travel is reality, and science fiction and futurism are fast becoming the immediate frontier. An understanding of both our historical and our physical relationship to our star will be the groundwork for our personal introduction to the universe.

Time, the Sun, and the Universe

When the sun ruled the pace of life, it was the basis of time. Separation from solar rhythms has altered both our individual and our collective world views throughout history because they are inextricably bound to an attitude toward time. For example, you may be the type of person who values your activities in terms of clock units, or perhaps you can't stand to be paced by a clock and prefer to judge your performance by successful completion of tasks. You aren't expected to clock in on a job if you're the vice-president, but you are often required to do so if you're a secretary. The Cree Indians did not even bother to count days on which they could not see the moon. Other ancestral communities saw time in recurring cycles of sowing and harvest or other seasonally repeated tasks, but contemporary industrial societies, on the other hand, view time as a marketable, consumable commodity to be sold in mathematical units. A mark of progress today is an increase in speed of production—a tool or process is "better" if it can make something more quickly and cost less time.

Having lost our sense of cyclic time, we've lost our identity within the space-time universe. Modern science, unlike that of the ancient world, has given us no dignified or emotionally satisfying images for the future. We have few adequate contemporary symbols to signify the *flow* of time, or continuous duration, and the cosmos. By contrast, the ancients had magic diagrams, geometric models, and mythology to illuminate, dramatize, and describe the indescribable universe. One of the simplest and most

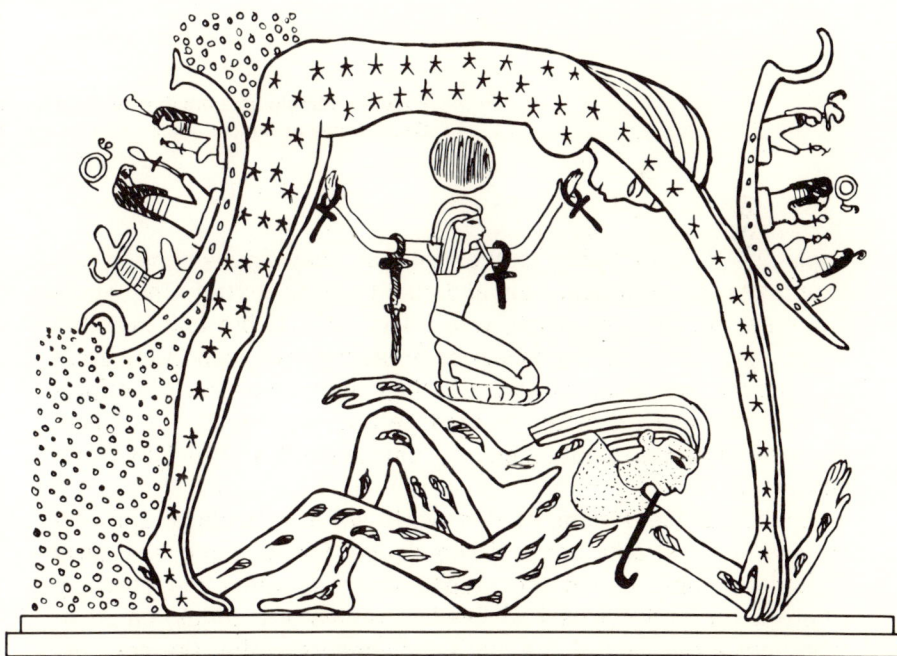

Egyption description of the indescribable universe. The sun god travels by boat over the back of the starry goddess each day and descends to the underworld each night.

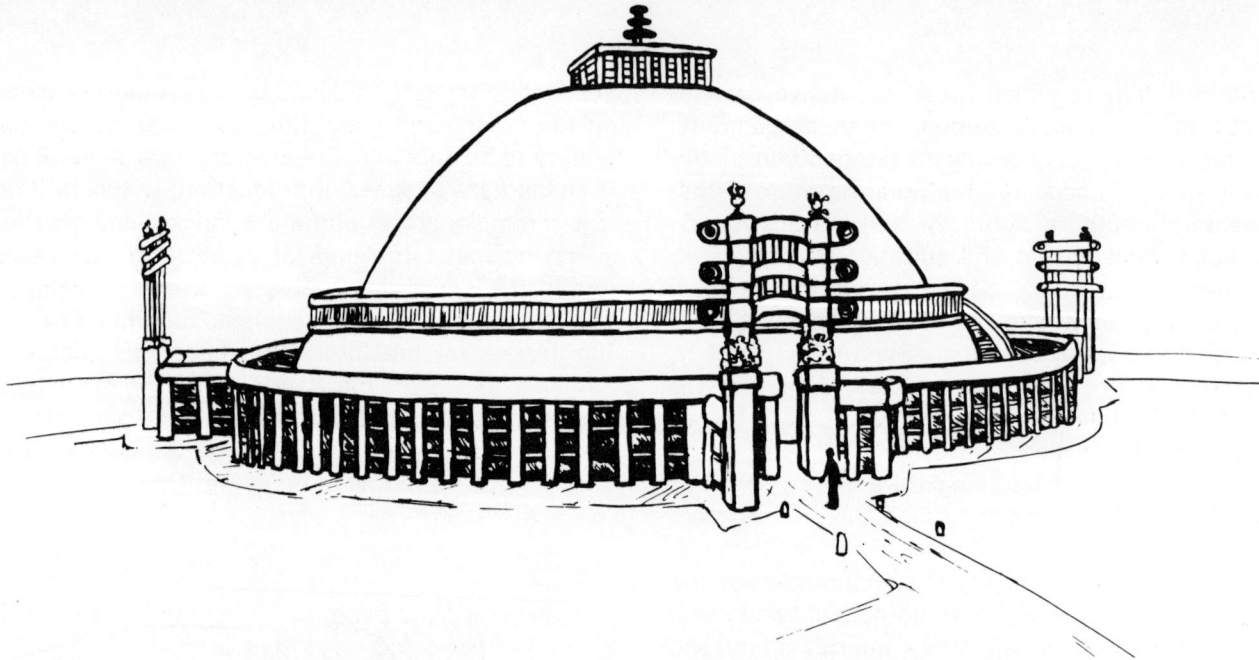

Buddhist temple as a meditative diagram of the sun's path and the cosmos (The Great Stupa at Sanchi, India, third century B.C. to first century A.D.).

commonly known today is the oriental *yin-yang* symbol, which expresses the continual flow and duality of all matter. The whole universe, not just the earth, was filled with visible and invisible beings and life forces, gods whose presence was seen in the stars and planets, and idealized forms in the celestial sphere. The efforts of religions were directed toward getting in contact with these elemental powers to obtain energy, physical and spiritual strength, and prosperity.

Visual representations of this world were symbols of spiritual illumination and astronomical knowledge. For example, the early Buddhist temple called the stupa, which signified the dwelling where souls and spirits had lived in prehistoric times, came to represent a diagram of the sun's movements within the cosmos. Its circular mound, called the World Mountain, was a metaphor for the sky or celestial sphere. The small three-part umbrellalike form on its top (a pointed spire on some temple versions) de-

picted several concepts: fire (sun symbol) of vision, the tree of life, the world axis, and the sanctuary beyond the earth. The four gates of the surrounding wall marked solar positions on the four corners of the celestial world (or true earth). The space between the outer wall and the stupa and the fenced balustrade on the mound itself were used for ritual walking in the direction of the sun's course. The entrance gates of the wall represented the sun's location during the course of a day: sunrise, zenith (the highest point overhead at noon), sunset, and nadir (the point opposite zenith, or midnight). The stupa reminded the worshipers that the sun lights up the physical world as Buddha illuminates the spiritual one. Offset entrance openings through the wall looked (from a bird's-eye aerial view) like the four arms of the swastika, another ancient sun symbol. Many variations of this basic structure were built between the third century B.C. and the twelfth century A.D. in Ceylon, India, and Nepal.

Unlike our ancestors, we have yet to learn how similarly to dramatize long stretches of time and thereby create images that illuminate our world. Perhaps we can start by heightening our awareness of the daily and seasonal passages, the patterns and quality of light from our sun at different times of the day and year. This might help us develop a sensitivity to the poetry of another time, which is cyclical. The sundial is uniquely suited for this task. It dramatizes the constantly kinetic diagrams of light and shadow that express the cosmic flow through the dial's direct link with the solar system.

Myth and the Sun

Myth was the language of science before and even after mathematics regulated our description of the universe. It was both a prelude to and the constant companion of our first science—astronomy—and recorded man's belief in cyclic time. Solar and astronomically based religions taught scientific theory through sacred performances, musical rites, and symbolic commands and gestures. Both Egyptian and early Indian priests walked around their temples in a circular path each day in order to suggest that the sun do the same. Teutonic recitation of the rune, an incantation that was sung or murmured, was supposed to create magic by invoking the origins of physical phenomena; for example, a rune about the heavenly origin of iron was supposed to heal a wound from a sword. (Iron was thought to come from the sky because the meteorites that were found were of iron.)

Myth, ritual, song, and dance were originally part of the seasonal celestial observances. Participants in the rites played the parts of gods and supernatural powers, and their group gestures became rhythmic ceremonial dances. It is claimed that all medieval folk dances were survivals of pagan imitative and generative magic. The dances of some American Indian tribes today, such as the corn and rain dances, are examples of symbolic commands. The accompanying tales and myths initiated the group into the workings of the universe in order to help individuals in a given community become more conscious agents of natural forces. The Navajos performed a circular sun dance, and the *quetzal* dancers of the Mexican Puebla highlands enacted a variation of it showing the relation of the sun to time. Because certain mystery rites forbade the spoken word, some lore could only be mimed or danced. Greek history records that a famous Greek mime was said to have expressed the Pythagorean philosophy in a brief dance.

Even children's rhymes and games today contain vestiges of celestial mythology. One scholar has claimed that "Hey Diddle Diddle, the cat and the fiddle, the cow jumped over the moon. The little dog laughed to see such sport, and the dish ran away with the spoon" was originally connected with Egyptian sky goddess Nut, who was a rain deity represented as a cow. The cat was Bast, or the star Sothis, and the fiddle was the instrument she played in a May festival. The dog was the god Anubis, and the dish and spoon were ritual containers. The Eskimos (Esquimaux of Igluik) played the string game of cat's cradle when the sun was progressing southward in the autumn in order to catch it in the mesh and so prevent its disappearance; after the winter solstice they switched to a game of cup and ball to speed the sun's return.

Today we celebrate our cyclic solar rites with Christmas and Easter. The date of Christmas was established to coincide with the winter solstice festival of the ancient Near Eastern sun goddess Mithra (known as Astarte in Semitic cultures). It was celebrated on December 25 of the Julian calendar throughout the Roman Empire in Africa, Egypt, Syria, Italy, Spain, Portugal, France, Germany, and Bulgaria. The worshipers went to shrines on the eve of the Nativity of the Sun and later burst out of them at midnight yelling: "The Virgin has brought forth!" The Egyptians even displayed an image of the newborn sun to the congregation. The early Christians of Egypt did not celebrate Jesus' birthday (the Bible said nothing about the date of his birth), but they established January 6 (the alleged baptism date) as the nativity by the end of the third century A.D. The Western Christian church never recognized the January date, nor did it celebrate any "birth" day, but chose December 25 to be Christmas (between A.D. 359 and 375) to compete with the pagan solar religion, and kept January 6 as the Feast of the Epiphany. Saint Augustine (A.D. 354–430) bade his Christian fellows "not to celebrate that day [December 25] on account of the sun, but on account of him who made the sun."

The original purpose of lighting the Christmas tree was to make a beacon to guide the sun god back—although decorating it with

Lighting the Christmas tree with sundials at the winter solstice to bring back the sun. *Photo by Rosmarie Hausherr.*

times a section was kept in a half-burnt condition to light the next year's Yule log. The Persians originated the custom of making a holiday fruitcake to be laid on sacred altars in honor of the sun—fruit trees were often pictorial accessories of the sun.

Many solar cultures had symbolic ball games, which represented the battle of the powers of the sun over darkness (although we don't know if they were played only at the winter solstice or not). The ritual Yule football (probably soccerlike) sports of the Scottish Highlands were called *luchd-vouil*, and the Welsh game *Bel Troed.* The Mayans of Mexico and Guatemala had large ball courts near their temples for the playing of *tlachtli.* This game represented the motions of the sun and moon and was played with a rubber ball that was half-light and half-dark. The patron god of the game, Xolotl, represented both light and shadow. He traveled with the sun and moon playing ball. Sometimes he won, other times he lost. The Hindu equivalent deity was Rahu, a sun-swallowing monster who caused eclipses as he pursued the sun and moon.

Another story, from the Chumash Indians of California, includes a ball game within a race.

The Tortoise and Hawk were to race around the globe on the day of Xuas [goddess of the earth], throwing balls ahead of themselves as they ran. The loser and his referee were to be burned alive. The animals of the world chose the Coyote as judge of the race. Then the Tortoise himself also selected the Coyote as his own referee— and by law the Coyote could not refuse. On the morning of the race the starting signal was given, and the Hawk took off. The Tortoise first struck his ball so hard the spectators could only see a streak of its flight, then he started crawling along. Meanwhile the Hawk encountered strong winds, which so completely disoriented and slowed him down that by the time he got near the finish line, he was worn out. The Tortoise, who kept moving steadily, was still energetic and crossed the line first. But the question of whose ball crossed the finish first was subject to great debate. The Coyote of course ruled in favor of the Tortoise—his life was also at stake in this decision. But the onlookers decided that the Hawk was the winner because the Tortoise did not finish at the same time as his ball. The Coyote ran away so quickly that the other animals couldn't

sundials offers a safer variation of the idea, lighting it with emblems of the sun. *Yuletide* comes from the medieval Anglo-Saxon *guil* or *iul*, meaning "sun." The Yule log represented the sun's body and so symbolically possessed the solar generative powers. Tradition had it that if you kept a charred fragment of its wood under your bed, it protected you and your house. If you put it in your plough, your sowed seeds would prosper. Some-

catch him. They didn't burn the Tortoise either because he said that he would urinate on the fire and put it out.

The tale is vague about the method used by the Hawk for moving his ball. It seems that each contestant had his own ball (possibly representing the sun and moon). The Tortoise hit his, but the Hawk seemed to carry it with him.

In Chumash folklore the Sun also played a kind of game, called *peon*, every night of the year with the inhabitants of the heavens. The Sun and his partner, Slo'w, a giant eagle who supported the heavens with his wings, played against the Coyote of the Sky (different from the Coyote of the Earth in the last story) and the Morning Star. The Moon was the referee. At the time of the winter solstice all points were totaled. If the Sun lost, it would be a rainy year with much food for the earth; if the Sun won, he took his prize in human lives. The Sun ate people.

Easter (from the Anglo-Saxon *eastre*, meaning "spring") was a solar rite of the early Roman Church that originated in the Greek and Roman feasts of Ceres (Demeter in Greek), the goddess of agriculture. She was associated with the seasons through her daughter, Proserpina, representing corn seed, who spent half the year in the underworld with her mate, Pluto, and reappeared on earth in the spring when she revisited her mother, Ceres. The Eleusinian Mysteries of the Greeks were secret rituals to honor Ceres, in which observers enacted scenes representing the seasonal alternation of life and death in nature. Roman fertility rites associated with Ceres involved setting animals on fire and driving them through the Circus, and in the Sada Festival of Persia burning animals were chased through the fields. Easter Eve fires used to be lit in Catholic countries. All lights in the churches were extinguished, and then the Paschal, or Easter, candle was relit with a new fire. In Germany a bonfire was also lit near the church, and its ashes were mingled with consecrated palm fronds to be mixed with seeds before sowing.

It is said that egg rolling was an Easter ball game of Christian ecclesiastics. A medieval drama called pace-egging had teams that engaged in mock combat until an adversary was mortally wounded and brought back to life. Another Easter ball game, similar to cricket, was played by both sexes with a dairymaid's stool as the wicket, and kisses and cakes were the winners' prizes. Both eggs and spheres were associated with fertility and the sun.

Another souvenir of solar myth is the dollar sign ($). This symbol was inscribed on the coat of arms of Charles V of Germany to represent the Pillars of Hercules and to illustrate his Latin motto *Plus Ultra* ("even farther"); it was later imprinted on German currency. Hercules, like many other heroes associated with light or the sun, was assigned many difficult tasks and travels around the world. He overcame obstacles and monsters representing both physical and spiritual darkness. During his journey around the globe he reached the boundary of Europe and Africa, which were at that time joined together at the area of Gibraltar. To commemorate his progress, he tore the mountain into two pieces—the Strait of Gibraltar separates them—which were called the Pillars of Hercules.

The Origin of Time Recording by the Sun

> Had we never seen the stars, and the sun, and the heavens, none of the words which we have spoken about the universe would ever have been uttered. But now the sight of day and night, and the months and the revolutions of the years have created number, and have given us a conception of time, and the power of inquiring about the nature of the universe; and from this source we have derived philosophy.
>
> Plato, *Timaeus* 47

Before the advent of clocks or calendars daily and seasonal events and tasks were anticipated by nature's cycles. Greek farmers in the time of Hesiod determined that plowing and sowing must be calculated by the departure and return of certain species of animals and plants; for example, sowing could be timed by the arrival of the cranes, vineyards had to be cultivated before snails climbed up the vines, and sea voyages could be safely made when the fig trees started to bud. Exactly which calendar dates these events correspond to varies, of course, by the particular climate of the location. Passage of the day was gauged by ancient Greeks and Malagasy in Madagascar by measuring their own shadows, with their feet as rulers. Rural settlements as late as the nineteenth century in Ireland used natural shadow-casting landmarks such as mountain peaks or tall trees or else set up piles of stones, called *cairns*, if there were no natural landmarks or suitable ground areas of great size to calibrate the shadows.

In many cultures, from ancient Egypt to colonial New England, midday marks were set by marking a line in the earth at the base of vertical poles. Labrador Indians on a hunt thrust sticks into the snow and drew an exact line of the shadow cast. The women who were following them with supplies knew by the advance of the shade away from these lines how recently their hunters had been there and approximately how far away they might be. Oriental and Near Eastern farmers set a line of *durra* (an Indian type of millet grain) stalk pegs into the earth, marking off a length between sunrise and sunset. The space between each peg was called an *alka*, from the Arabic word meaning "to hang" or "to hitch on." When the shadow of one peg reached the next, the farmer "hitched on" or changed oxen on the plows. All these timekeepers were devised naturally, without the help of accompanying mathematical theories and records.

Only after man began recording these daily observations did he use the clearly delineated patterns to suggest scientific theories of how it all worked. Writing, measuring devices, and numbers presented the opportunity to turn observation into abstract reasoning. Astronomy and philosophy were born—and the midwife was the mathematically constructed sundial.

Science Starts with the Sundial

The sundial was the first instrument of science. By *science*, we mean the kind of analytic study that is based on abstract order—that is, counting, measurement, and exact record keeping. Sundials are any objects precisely marked at intervals that use sunlight or shadow to give information about time, our planetary and stellar motions, or our location on this globe.

Civilization's consciousness of time started with certain realizations about the moving shadow and light cast by all sundials—natural and man-made. The sundial provided a view of a four-dimensional world (three dimensions plus time) and the idea that the cosmos is a moving flux. Fascination with this kinetic image stimulated wonder and religious awe, which augmented daily existence. In addition, the dial's development paralleled systematic philosophical reflection upon a complex universe and was the inspiration for early research into astronomy. Sundials were used to describe not merely the time of day (or time of year) but also other celestial phenomena. For example, the Neolithic observatory at Stonehenge, England, which was composed of upright stone megaliths set in a circle, was not only a giant solar calendar or yearly sundial; it also predicted eclipses of the sun and moon. Its rocks were aligned so that sunrises, moonrises, sunsets, and moonsets were framed between the upright stones, and an observer could sight along them out to the sky. This dial used the position of the sun's lighted disc rising and setting over the stones as an indicator of time.

In the third century B.C., Eratosthenes measured shadows cast by upright poles to show that the earth was round, and in the process he also determined its circumference. Hipparchus (ca. 140 B.C.) rediscovered the precession of the equinoxes (understood in Mesopotamia many centuries earlier) by using the first recorded model of the celestial globe (later called the armillary sphere, which we will discuss in chapter 5) to calculate his astronomical theories even before this instrument was used as a sundial.

The astrolabe, an instrument used for navigation invented by the Arabs about 150 B.C., was also a type of sun-clock, for it recorded the time, the date as represented by the sun's position in the zodiac, the latitude, and the angle of elevation of celestial bodies. If one knew any two of these, a third could be calculated.

You may make a sundial on almost any sunlit surface (flat, concave, or convex) in any position (horizontal, vertical, and at angles in between), even for night reading by moonlight (which is, of course, reflected sunlight), and also for constellations rising over points marked on the instrument. Renaissance dial virtuosos with secret construction methods competed with each other to produce sunken, heart-shaped, cylindrical, and many-faceted versions. Sundials were used as temple offerings in pre-Christian times, ritual objects in ancient China, and were considered to be the proper gift to present to kings and queens after the Renaissance. The dials made reference to the amazing synchronized mechanism of the whole celestial system and to the ideal eternal forms over which the Greek philosophers had continually rhapsodized. When the sun was god, the sundial was his scepter.

The Solar Connection

The sundial is an instrument bridging past and future, earth and outer space, nature and technology. It helps us encounter the future with knowledge and understanding because the sundial is the most immediately accessible means of charting the sun's trajectory and also our own geographic relationship to the sun from our specific location on this planet.

Sundials humanize and personalize our own time and make it special. Daily observance of light and shadow sweeping across the dials in geometric patterns gives us a heightened perception of the passing seasons and offers a look at another world system beyond that of the clock. Sundials bring visual poetry into our lives and help us to develop an individual sense of cyclic time— the awareness that phenomena and conditions repeat themselves in time and in space as well, so that similar events are often going on simultaneously in different areas of the world. (The telephone is said to have been invented in Russia at the same time that Alexander Bell developed it in America.) And all of these cycles may be related to the sun, either directly or indirectly.

This cyclic view of time is the basis of mythology and some science fiction and encompasses a cosmic world view that, sadly, we lack today. In mythical thinking, time is essentially a method of dealing with the flow we feel. Myth treats time like space by organizing it into areas that are special or sacred, profane or common, lucky or unlucky. Each time and season has a distinct and personal quality. Even by invoking the magic phrase *once upon a time* at the start of a tale one places the story in a special time-space apart from the ordinary. The biblical text of Ecclesiastes reminds us that there is a personalized time-place for every person and task under heaven: a time to reap, a time to sow, a time for happiness, and a time for sorrow.

A cyclic event becomes poignantly and emotionally charged by merging the past and future into the present. On New Years' Eve we automatically recollect other New Year's celebrations and speculate on future ones, more so than on all the other nonspecial days. When we wish someone "Happy New Year!" we express the cyclic belief and hope that inspires us to make new resolutions to change our lives.

When we deal with time this way, we compress and juxtapose events much the way films and novels do, infusing them with creativity. It is no coincidence that this cyclic view of time is represented throughout the Orient by the image of the snake biting its tail forming a circle (the *ouroboros*), which represents the path of the sun, the unceasing circular motion of the universe, and eternal time. The Kundalini Yoga of India uses the snake as a metaphor for creative forces and energies that travel up the spinal marrow in the body through *chakras*, or centers of psychic force.

You can share in this ritual tradition of mythical time and unbounded creativity by following the pilgrimages of the sun on the miniature models in this book. They have their origins in the study of man's reliance on the sun. Some of the cutouts are based on variations of historic dials; others are astronomical models of instruments to measure your location—not only to help you find out where you are on your planet but also to enable you to test your dials accurately and set them up properly. All are intended to reveal the dramatic poetry of our star's majestic radiance. We invite you to play in the sun.

Leni-Lenape, American Indian (the sun, moon, and stars)

Oriental sun symbol
(rising sun)

II
General Instructions for Assembly of Models

The following instructions list the basic tools and materials as well as describe helpful techniques for assembling your cutouts.

Tools and Materials

BASIC

1. Sharp scissors
2. X-Acto knife or a single-edged safety razor blade
3. Metal-edged calibrated ruler
4. Elmer's, Sobo, or any other white or clear model glue
5. Masking tape to hold parts while glue is drying
6. Pencil
7. Ponce wheel (also known as a dressmaker's pattern-tracing wheel) or a dull round-tipped tool like a butter knife (without teeth)

ADDITIONAL OPTIONAL MATERIALS

8. Coloring media

General Procedures

1. Take your time and follow instructions in the order presented.*
2. Tear out the whole page containing the plate pattern you are going to assemble. This will make the pattern easier to handle and will also insure that you do not cut the pattern following it.
3. When you cut with an X-Acto or other blade, protect the work surface below your pattern with an old magazine or piece of heavy cardboard.
4. Either cut lines first on the gray lines, or score lines first on the dashed (---) or dotted (...) lines. The order will vary according to the construction of the dial.

 * Note: These instructions have been written and ordered on the assumption that you will be using an X-Acto or other sharp blade (which we use for the best and most precise results). We often tell you to cut out slits and holes before cutting out the entire piece because small parts will move around when you cut them with a blade. If you use scissors instead for cutting, you may find it more convenient to change the order of instructions and cut out entire pieces first (which you can hold while you cut tiny sections). However, we *strongly* recommend that you use a sharp blade for cutting out very small or intricate pieces, or your end results may be disappointing.

Scoring Techniques

1. For the cleanest, most precise folds, score the fold lines before folding them.
2. Score lines by rolling a ponce wheel (pattern-tracing wheel) or other scoring tool along them. Use a straightedge or ruler as a guide on straight lines.
3. Always fold paper back *away* from the scored lines.
4. Score all dashed lines (___) on the printed, or top, side of the page and all dotted lines (...) on the blank side, or underside, of the page, and be careful not to cut through the paper.
5. When score lines must be made on the blank side of the page, poke a tiny pinhole through the last score dot at each end of that line on the printed side of the page. Then when you turn the page over in order to score the blank reverse side, you can connect these pinhole guide points with your score line.

Additional Points of Reference

1. When we speak of the sun moving across the sky, we are referring to "apparent" motion. In reality we all know that the earth is moving relative to the sun, not vice versa.
2. When we give instructions to position or align your models so they point south or north, we are doing so in relation to our own position in the northern hemisphere. In the southern hemisphere, the orientation of the dial should be reversed.

Some Additional Options

These assembled cutouts may be used as they are, placed on windowsills, as original artworks placed on shelves or the Christmas tree, or sent as greeting cards. If you wish to use the models outdoors (for example, at the beach to time your suntan), spray them with plastic waterproofing in order to make them more durable. They can also serve as patterns for making more copies in permanent materials, such as sheet metal or plastic (acrylic for flat parts and heavy flexible vinyl for curved ones). You can also curve rigid acrylic plastic sheets by heating them with a hair dryer or on low heat in an oven and then bending them while warm; however, this technique of heat forming takes a bit of skill and experience. You may increase the size of your models by tracing the patterns on a grid. However, remember also to increase the thickness of the material they are cut from (and, of course, the slots and holes also), or the forms will look weak and insubstantial.

Transferring Models to Other Materials

Combinations of materials can be used for decorative effects as long as you account for their thicknesses when making slots, holes, and the like. For example, if you use a dowel or metal rod thicker than the dimension called for in the original pattern, you will have to make an alteration on the part it fits into. In addition, the shadow cast on the dial face will also be thicker or broader, and you will have to read the time from the center point of the shadow-casting stick. Heavy leather might also be embossed with gold leaf hour marks for a horizontal dial face.

Coloring Your Models

Transparent media such as Magic Marker and colored pencils are preferred (unless you are neat and careful about staying inside the lines) because they will not obscure the numbers and other markings on the dials. Opaque media such as crayons, tempera, enamels, model paints, and the like are suitable for parts without numbers. Apply wet colorants very sparingly, or else the paper of your dials will warp. If you do use opaque paint on parts you are copying, be sure to paint or spray the color on first and add the numbers on top afterward, using carbon or transfer paper to get correct positioning of the marks. When you use spray can colors, *always* have excellent ventilation. They are toxic and might make you sick otherwise.

We hope you come to delight in the sensuous traits of paper: its direct responsiveness to quick modeling of angle, plane, and curve; its flexibility, colorability, and even its disposability. All these traits make it a medium of play. When you've mastered your original model, you can either use it to invent new variations or to trace another pattern. Part of the fun here is that you can take artful risks to make your own variations in the painting. You can even alter the sun sculptures. So what if you can't always tell every hour on it anymore? Or if you're extra careful about taking many of the pieces out of the heavy paper at the back of the book with an X-Acto blade, instead of cutting into the sheets from the margin edges with scissors, you may be able to use parts of that page again and again as a stencil by slipping another piece of paper under the cut-out holes left in it. This method won't always work, but you can also reinforce previously cut areas of the pages with cellophane or masking tape.

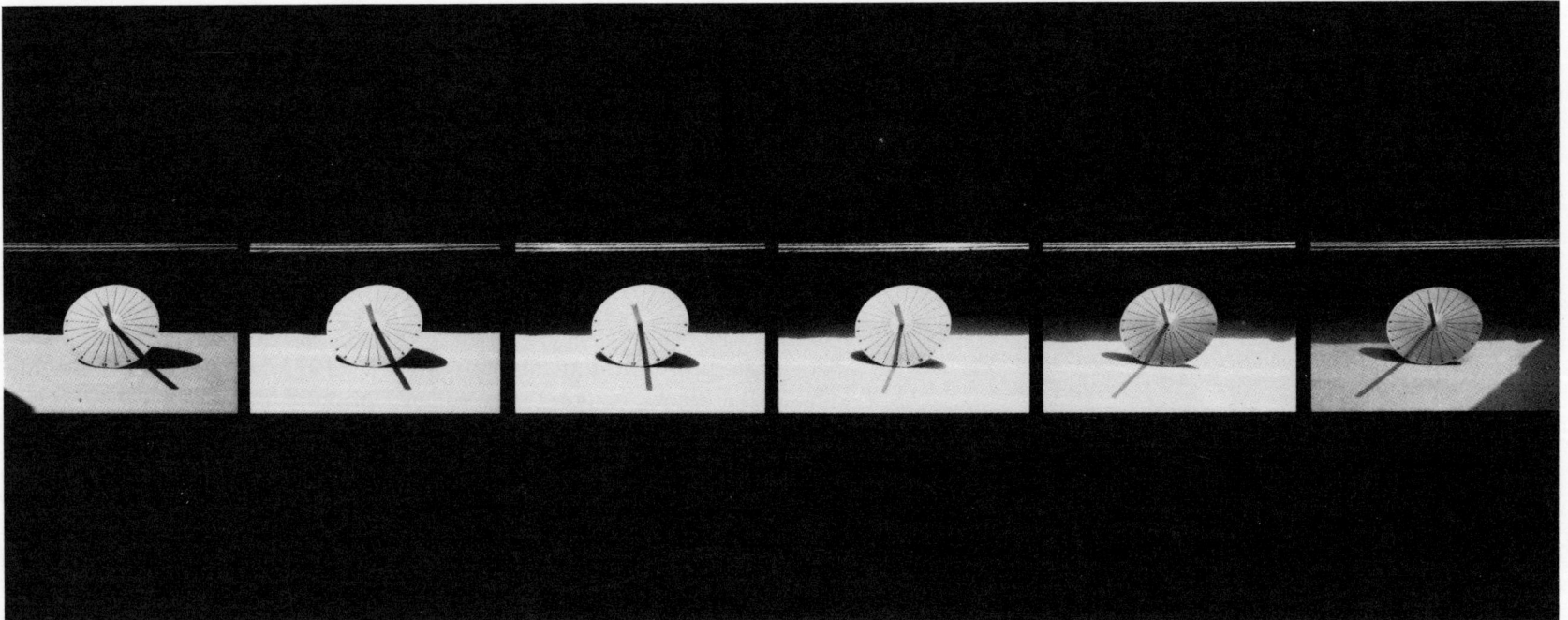

Model 1. "The Great Wheel" (an equatorial sundial), showing different hours.
Photo by Rosmarie Hausherr.

III
"The Great Wheel"—The Equatorial Sundial

Our expression "the great wheel of time" was initially a literal description of the grand mythical image of the sun, time, and the heavenly cosmos. The sun wheel was a universal symbol from prehistoric times onward. Its exact meaning was not precisely the same for everybody because the image represented ideas as well as an object, which left a great deal of room for different interpretations. Two-dimensional drawings of sun wheels depicted not only the solar orb but also the revolving universe, eternity, and fertility because the sun was seen as the source of all life, energy, and power. The rays of the wheel spokes represented the four cardinal directions for the American Indian and the elements and mysteries for the Chinese. Ancient Greeks saw the wheel as part of a golden chariot, and some Hindu temples symbolizing the sun either had wheels carved on their sides (for example, the Black Pagoda of Konarak, dedicated to the sun god about A.D. 1240) or actually were movable chariotlike shrines called Juggernauts (or *Jagannaths*), which were pulled by hundreds of religious followers through the streets once a year. Devotees threw themselves under the Juggernaut wheels as a shortcut to paradise until British colonial law ended the practice. Early Buddhists expected a jeweled wheel with a thousand rays to appear in the heavens, radiating spiritual light as the sun did physical light,

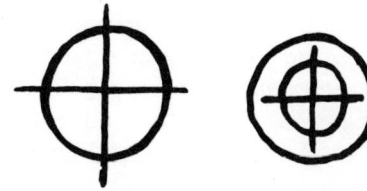

Stone Age Scandinavia

Lono, Congo

Lono, Congo

Stone Age England

American Plains Indian

Pre-Columbian Mexican

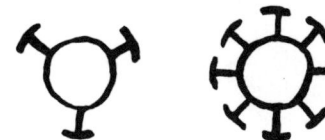

Greek

Examples of sun wheel diagrams

when an enlightened virtuous ruler came to set up a reign of righteousness and good law throughout the world. The swastika symbol is derived from the sun wheel, and it was seen as a solar symbol of good luck. The rays may go clockwise or counterclockwise, respectively representing the masculine turning outward or the feminine turning toward the inner self. Hitler very deliberately chose this ancient form as a symbol of power for his regime.

The three-dimensional wheel was also a model of the physical universe and described how it moved. The circular plane of the wheel divided the sky into two halves and turned on a central mythical axle, which was variously called a world pillar (in Palestine at the time of the biblical story of Samson), a churning staff (by the Maya in Mexico), an oak tree of life (by Celtic Druids), or other equivalents. This axle pointed to the north polar star, which for us today is Polaris. In many other areas, such as ancient Egypt, Scandinavia, and Iceland, the wheel was interpreted as a millstone grinding out the events of time. Hindu Indians saw it as a wheel of fire with a fire drill stick (for kindling the flames) in the center. Plato spoke of it in *The Republic* as the spindle of necessity that was held by the fates, who were spinning and unwinding the threads of our lives. The turning of this celestial wheel, caused by the passage of the apparent sun through the signs of the zodiac in the sky, determined earthly events in a series of epochs governed by different constellations.

Your model of the mythical cosmic wheel here is the simplest form of sundial. Cut out and assemble plate 1 according to the instructions.

"THE GREAT WHEEL" (AN EQUATORIAL SUNDIAL)

GENERAL INSTRUCTIONS
1. Remove plate 1 from the back of the book before cutting out parts.
2. Cut only on solid gray lines.
3. On the printed, or top, side of the page score all broken lines *before* cutting parts from the page. Score lines with either a pattern-tracing wheel (also called a ponce wheel) or with a dull, slightly rounded tool, such as a butter knife (without teeth).

4. Fold all scored lines back away from the side of the paper containing the score.
5. Use a straightedge as a guide for cutting (with an X-Acto blade or mat knife), scoring, and folding all straight lines on the shadow-casting stick (called a *gnomon*) B for the most precise results.

ASSEMBLY
1. Score A and B according to general instructions.
2. Cut out and fold parts.
   a. Cut out A, and cut on crossed lines inside central square. Fold down central triangular tabs as shown in illustration 1.

   b. Cut out B and trim off calibrated end at the line equal to your latitude number. For example, if you live at 40° north latitude, trim at 40. (See p. 26, Positioning Your Dials.) Fold B on score lines to form a square tube. Glue tabs inside the end and hold tube closed with tape until glue dries. See illustration 2.
   c. Cut out right triangle C.

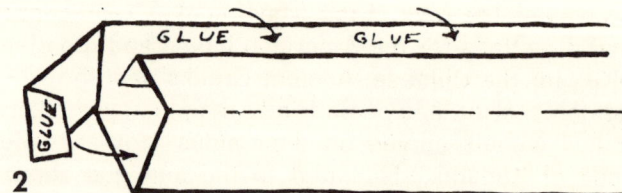

3. Join parts together.
   a. Insert the cut-off end of B through square hole on the printed or top side of A until A reaches the 90° mark on B.
   b. Glue A to B, placing glue on triangular center tabs of A. Be careful to keep A and B at right angles to each other by using right triangle

Model 1. "The Great Wheel." *Photo by Rosmarie Hausherr.*

C to check the angles all around B. See illustration 3. While glue is drying, set assembled model over an open container, such as a can, glass, or box, to keep it correctly aligned. See illustration 4.

HOW TO USE YOUR WHEEL

1. Point closed end of gnomon B upward and north and the open, cut-off end on the ground toward the south.
2. Position the twelve o'clock mark on A so it rests on the ground at the center of the bottom, flattened edge.
3. To set dial rotate disc A until shadow of gnomon B falls on correct hour. See illustration 5.
4. In the summer the shadow will be cast on the upper (or printed) side of the dial, and in the winter it will be on the under (or blank) side. Because the paper here is transclucent, your shadow on this model will shine through the paper in the winter so you can still read the printed side then. If you wish to make an opaque copy of this dial, mark the hour numbers on the underside as well for winter viewing.

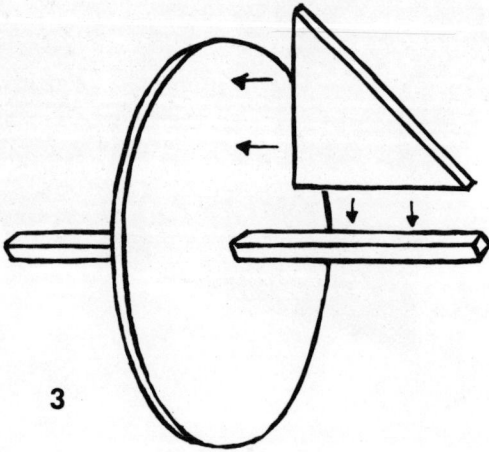

3

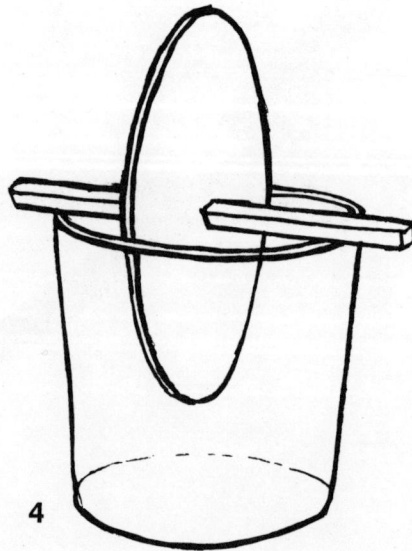

4

5

TO NORTH POLE STAR

Since the principle of this equatorial sundial is so simple that it can be made without complicated mathematical calculations and can also be used all over the world, you may wonder why it is not as common as the ordinary variety of horizontal dial with the triangular-shaped gnomon, probably the type with which you're most familiar from gardens and backyards.

Look again at your equatorial wheel to see exactly where the shadow is being cast. If you are observing it during the months of October through March (more precisely from September 24 to March 20), the shadow is on the under or bottom side because the path of the sun dips low in the winter sky below the surface of the circular disc. The round plane of the dial represents our earth's equatorial plane.

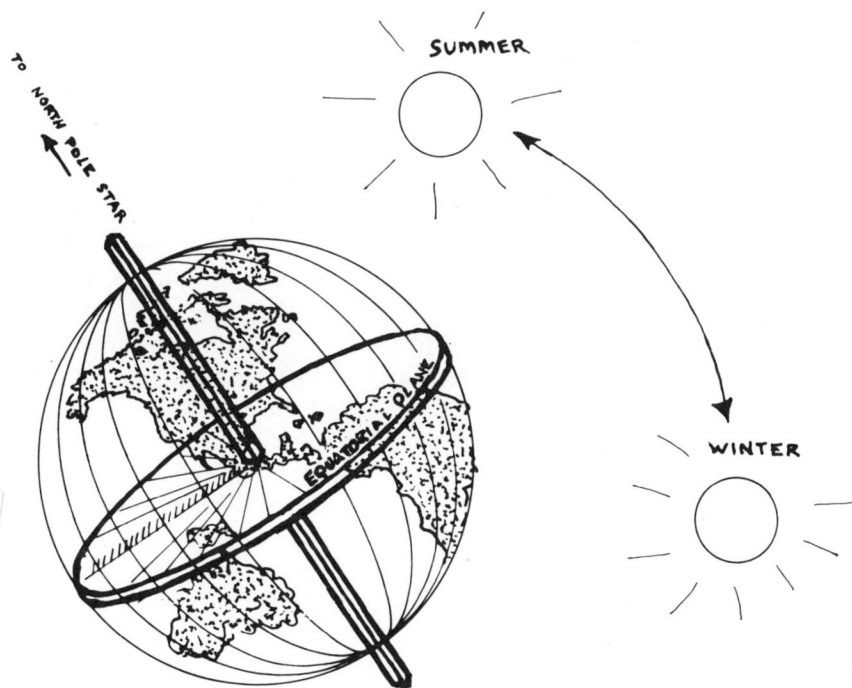

The relationship of the equatorial sundial to the earth's equator (or equatorial plane).

Actually we do not see more of these dials because of a timing problem. An opaque wheel would have to be placed high up on a tall pedestal above eye level or on a building so that one could see its underside in the winter, or else it would have to be translucent for public viewing all year around. The common horizontal types of dials are easy to read in all seasons because their shadows are cast on flat tables and are visible all year round. However, some knowledge of mathematics is required to construct them. You will make your own horizontal dial later in the book with the math already solved for you.

Look at your model again. If you turn your center, shadow-casting pole, or axle, called a *gnomon* (pronounced nó-man), straight up vertically so that the circular disc represents the ground, you have a kind of obelisk. Egyptian obelisks (about 1450 B.C.), which were four-sided stone pillars tapering up to a pyramid-shaped top, were placed next to temple walls as symbols of solar rays and were not used for telling time. "Cleopatra's Needle," an obelisk in Central Park in New York City, is actually a transplanted Egyptian original (now unfortunately eroding from the city's air pollution, although it was thousands of years old when it came to New York). The Romans took the Egyptian obelisks to use as sundials, not knowing their original religious function. Two basic form ideas in both Egypt and India were the ground-hugging square and rectangle, used to represent terrestrial sanctuaries, and the upright shaft pointing to the sky as a celestial image. The grounded forms were often associated with lunar worship and the vertical ones with solar worship. The pyramid is anchored visually to the earth but it also aspires upward toward the sun, representing the best of both worlds. The Indian temple form called the *stupa* also combined lunar and solar principles.

We did not make a pyramid top on your miniature wheel axle here because it is difficult to make that tiny pointed form on this small scale. However, if you like challenges, you might try making a pointed pyramid top yourself on this or on a larger copy of your wheel.

As you turn your gnomon to this straight-up position and mark the locations of the cast shadow every hour by your watch, you will notice that these marks do *not* fall on the regularly spaced marked intervals and may even fail to stay on your wheel at times (depending on your latitude). On an obelisk, or vertical gnomon, the shadow of a perpendicular upright moves faster in the morning and evening than around midday because of the oblique angle of the sun at those times. The wheel in this position will only indicate unequal temporal hours, a time distinction that was considered perfectly natural until the Renaissance. Since the hours of daylight and night are unequal except during the spring and fall equinoxes, there is no equal amount of light on successive days, and the hourly intervals themselves are unequal on the obelisk-type dial. Agricultural people using only daylight for work did not think of dividing their time in any other way. Only the Chinese and astronomers used equal time intervals (in the Roman Empire they were called equinoctial hours), which were measured by the stars.

In the fifteenth century, two centuries after mechanical clocks were popularly introduced, civilization gradually began to accept the idea of equal hour divisions. The use of temporal hours persisted in Japan until the nineteenth century, and the dials of mechanical clocks were adapted to the unequal system. When the ancient Egyptians used the water clock, a device which dripped at a constant rate, at night and on cloudy days, they took great pains to calibrate the water flow irregularly so that it agreed with their sacred cosmic timekeeper—the sun. In its simplest form, the water clock, or *clepsydra*, was a bowl with a hole in its bottom set in a larger container of water. When the smaller vessel filled and sank within the larger vessel, one time interval had passed. It is interesting to note that the length of this interval varied, depending on the season, temperature, and air pressure. This clock was used to time the speeches of politicians in Greece and Rome. Other elaborate versions were invented as far away as China, but of course they all had the disadvantage in northern countries that they froze solid in the winter.

The historical transitions in society that accompanied the change to the equal-hour clock system were profuse and complex.

"Cleopatra's Needle," Egyptian obelisk inscribed with hieroglyphics made for Pharaoh Thothmes III in the sixteenth century B.C. in Heliopolis. It was moved to Alexandria, Egypt, in 12 B.C. by the Romans and was a gift from Egypt to New York City in 1881.

We have calibrated the sundials in this book on equal clock hours in deference to modern-day thinking.

Your "Great Wheel" instructions and accompanying illustration (in the section "How to Use Your Wheel") tell you to tilt your gnomon at an angle to the earth equal to your own latitude. When the wheel is in position, it gives you the equal-hour intervals. The wheel takes its generic name—equatorial sundial—from the position in which it is used. Its flat disc is aligned parallel to our earth's equator so that the perpendicular gnomon points toward the north polar star.

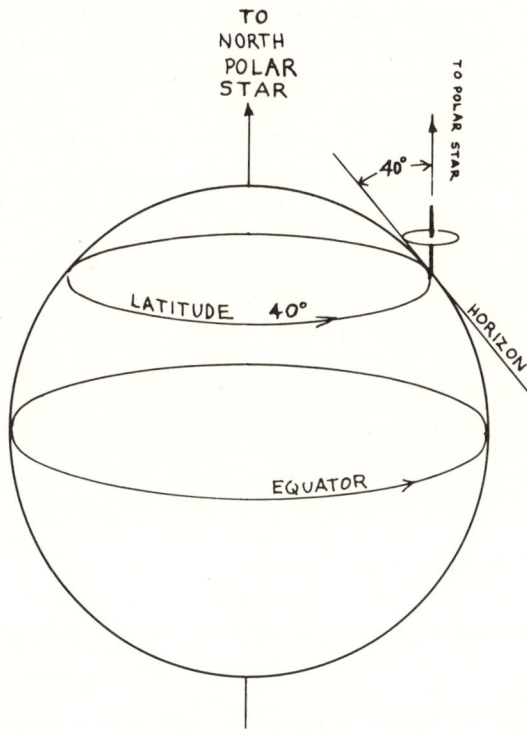

To understand how it works imagine that the axle of this wheel is the earth's axis rotating around a full 360° in twenty-four hours, or 15° per hour. So twenty-four equally spaced divisions on this circle each represent an hour. However, since much of this time is at night, we only number the daylight hours on the disc.

This wheel is not only the basis for many later portable dials carried by early travelers, but it is in fact the basis from which most contemporary sundials are derived and developed. The Chinese used this type of dial by the third century B.C., for they did not accept the Western temporal hours. By the twelfth century A.D. it was carried to the West, possibly by the Arabs or Jews. Although the exact European invention of this dial is obscured in prehistory, walls of some ancient buildings in England were incised with little circles containing lines radiating from a hole or indentation in their centers (which may have held gnomons). The earliest ones, with very shallow center holes, may have been symbolic sun wheels, since they were often in out-of-the-sun places under roof eaves and in shaded areas. However, by the late seventh century A.D. these circles began to appear on sunny southern church walls and were called Saxon dials, scratch dials (because they were scratched on the walls), or mass clocks, and it is supposed that they marked prayer times.

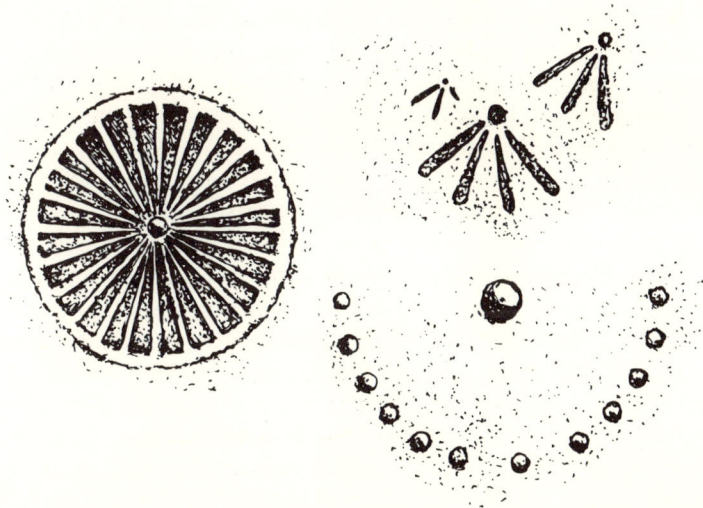

Saxon dials (also called scratch dials and mass clocks). These wheel shapes were scratched onto vertical sunlit walls in early England and were thought to have held shadow-casting gnomons in their center holes.

The placement of their radial lines was often so inconsistent that some scholars question whether they were actually sundials. But unless these rural people lined up their walls precisely along the true north–south meridian line and accurately tilted their gnomons for their latitudes, some inconsistencies should have been expected.

Positioning Your Dials

Two essential elements will allow you correctly to position any of your dials: You must find out your latitude and where true north is. There are several methods by which you can do this. The first is a shortcut requiring no understanding of how the dial functions or any further paper construction. The other method requires making two more cutouts and will increase your understanding of solar movements.

THE QUICK METHOD

Find your latitude by consulting a good map or almanac or by phoning your county surveyor, city engineer, or local museum of science or planetarium. If you use a map, every 69 miles (or 111 kilometers) equals 1° of latitude. To set your dial at its latitude mark, find a nearby city with a marked latitude and then add 1° of latitude for every 69 miles you are north of it, or subtract 1° if you are south of it. If you live south of the equator, add if you are south of the city and subtract if you are north.

Next, find the approximate true north–south line by using a watch. Place the assembled paper dial in sunlight at noon (Standard Time) with its gnomon in the general direction of north (or south if its instructions specifically call for that) and turn it until the shadow cast on the marked table falls on the number indicated by your watch. You can expect that the dial may be as much as sixteen minutes fast or slow, depending on the season, because at different times of the year the earth moves at different speeds.

MAKING MODELS TO FIND NORTH–SOUTH LINE AND LATITUDE

"Hindu Circles": The Method of Using Equal Altitudes

Although you can also set up your dials by several other methods, using additional instruments, we have included the next model for finding true directions in a fun and simple way.

P.S.—A magnetic compass tells magnetic, not celestial—or true—north, and you must make mathematical corrections to find the true direction.

You could merely place an upright stick in the ground and use a piece of string to measure out equal circles and then bisect the arc with it (as did many peoples through the centuries, including early Hindus in India), but we prefer to use a more elegant paper cutout version. The basic method of equal altitudes was described in Chinese scrolls in the second century B.C. Locating directions was important for various reasons: For instance, it situated travelers in unknown territories, and it also showed Muslims the direction of Mecca—something they had to know in order to face it while praying.

The direction-finding method we will use consists of marking a horizontal paper surface during the morning when the light (or shadow in other versions of this device) falls on any one of the concentric circles on it and then marking it in the afternoon when it again falls on the same line in a different place.

Cut out and assemble plate 2.

"HINDU CIRCLES"

(to Locate True Directions by the Equal Altitudes Method)

ASSEMBLY

1. Remove plate 2 from the back of the book before cutting out parts.
   Note! This entire sheet is used. Do not cut out and detach any individual sections from the plate except for the center triangle.
2. Cut only on gray lines. See illustration 1.

SOUTH

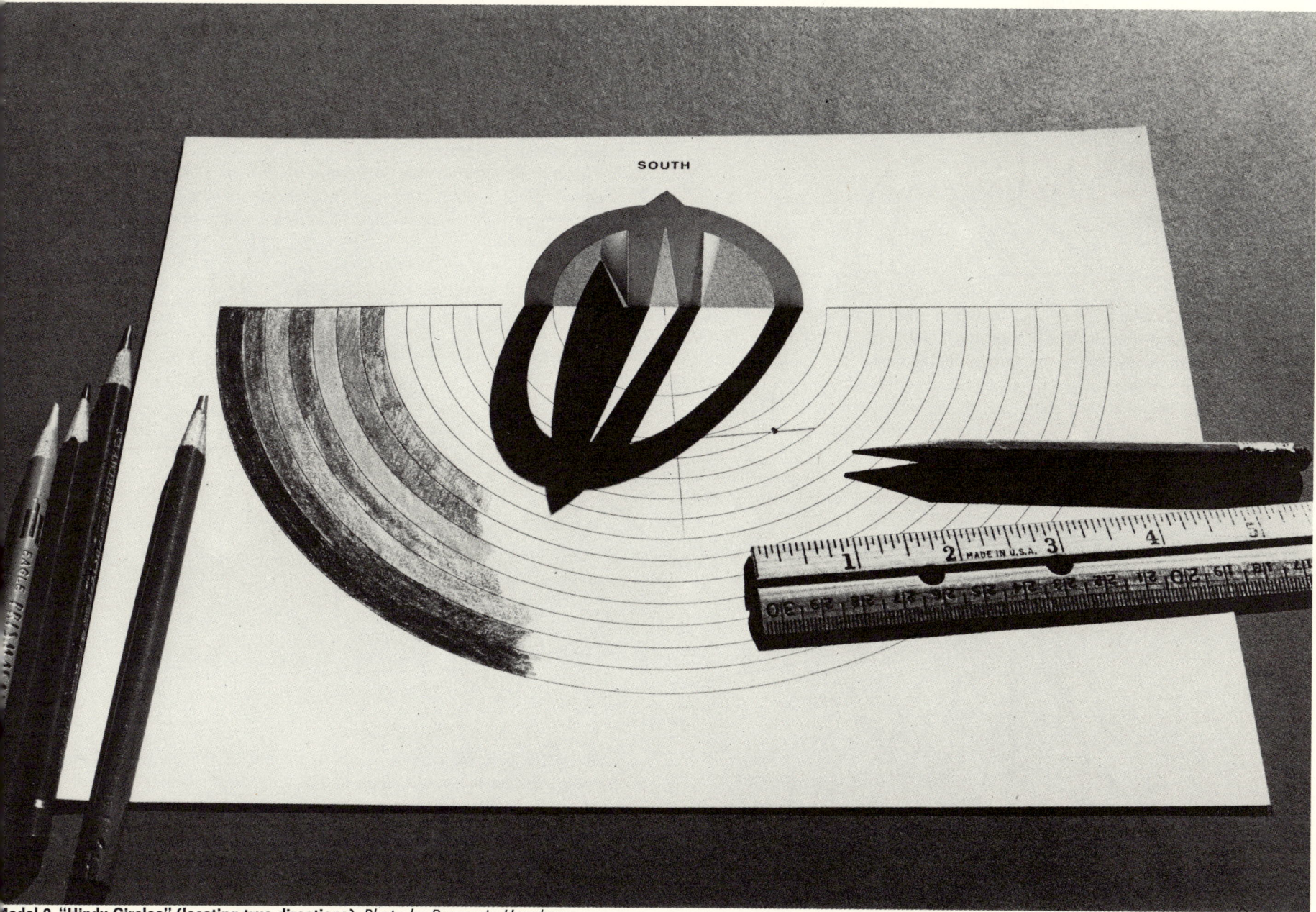

Model 2. "Hindu Circles" (locating true directions). *Photo by Rosmarie Hausherr.*

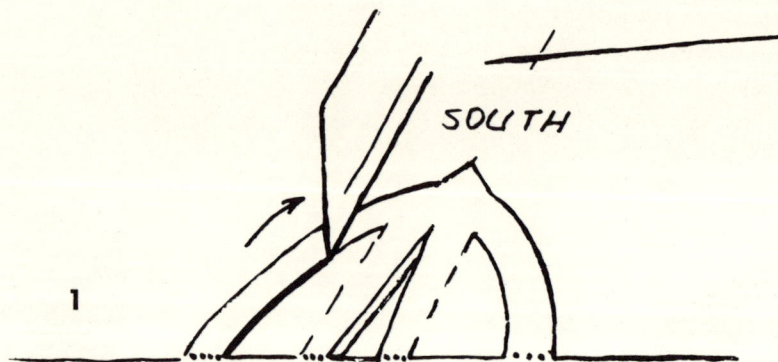

1

3. Score all dashed lines (_ _ _) on the printed, or top, side of the page. Score all dotted lines (. . .) on the blank side, or underside, of the page. Be careful not to cut through the paper. Score lines with either a pattern-tracing wheel (also called a ponce wheel) or with a dull, slightly rounded tool, such as a butter knife (without teeth).
4. Fold all scored lines back away from the side of the paper containing the score.
5. Use a straightedge as a guide for cutting (with an X-Acto blade or mat knife), scoring, and folding all straight lines for the most precise results.
6. Place the page on a level surface and fold up the arch. Prop it upright by opening back the inner flaps. See illustration 2.

2

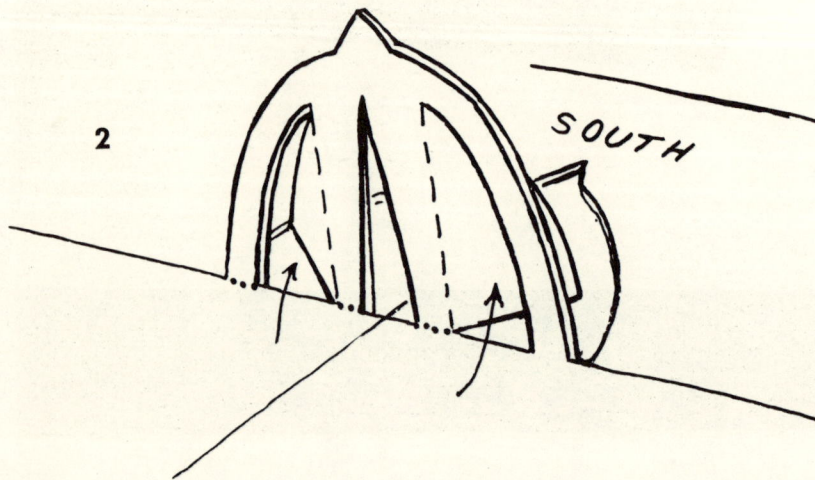

HOW TO USE THIS MODEL
1. Place model on a level surface so the south mark (the triangular cutout arrow on the ground plane in back of the upright arch) faces the general direction of south.* Secure the paper with tape or tacks to the surface you are working on so it will not move while you are taking measurements.
2. In the morning mark point A where the tip of the *sunlight* (not shadow) triangle shines through the center opening of the foldup and crosses one of the concentric half circles on the horizontal plane. See illustration 3.

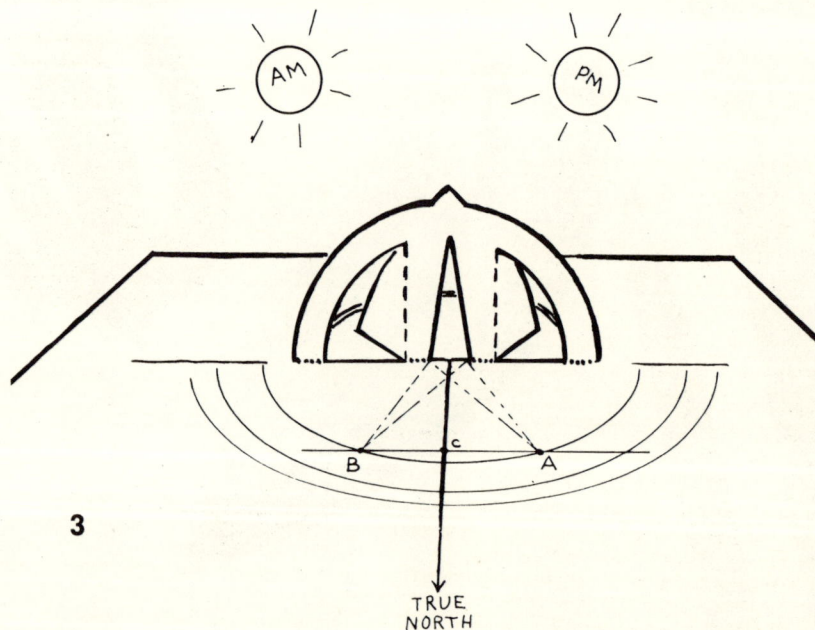

3

3. In the afternoon mark point B where the tip of the same sunlight triangle crosses that *same* circle line at a different place. (The angle of the sun will then be at an altitude equal to that of the morning mark.)

* It should face north if you live in the southern hemisphere.

4. Connect A and B with a straight line. Next use your ruler to bisect that line in its center with a dot (which we call C in illustration 3).
5. Draw a straight line from C to the center point of your foldup. This is your true north–south meridian line, which will be used repeatedly to align your noon mark on your sundial. If you mark this line permanently on your windowsill, you can now remove and frame your "Hindu Circles" for hanging on the wall.

The Latitude Finder

If you would also like to find your latitude with another paper model, cut out and assemble plate 3, which we call a latitude finder. This form is derived from the astrolabe, the astronomical and navigational instrument described earlier, which was invented by the Arabs around 150 B.C. Here we have taken inspiration not from the main body of the tool itself but from one of its accessories—a scale for measuring shadows to calculate the height, or altitude, of the sun. Then we consult a chart of the sun's known seasonal positions at different times to find our latitude.

LATITUDE FINDER

GENERAL INSTRUCTIONS
1. Remove plate 3 from the back of the book before cutting out parts.
2. Cut only on gray lines.

Model 3. Latitude finder. *Photo by Rosmarie Hausherr.*

3. Score all dashed lines (___) on the printed, or top, side of the page. Score all dotted lines (...) on the blank side, or underside, of the page. Be careful not to cut through the paper. Score lines with either a pattern-tracing wheel (also called a ponce wheel) or with a dull, slightly rounded tool, such as a butter knife (without teeth).
4. Fold all scored lines back away from the side of the paper containing the score.
5. Use a straightedge as a guide for cutting (with an X-Acto blade or mat knife), scoring, and folding all straight lines for the most precise results.

ASSEMBLY
1. Cut out A. Next carefully cut out slots on center line with an X-Acto blade or single-edged razor blade. Then cut the inner half-circle arm.
2. Score A according to general instructions.
3. Cut out B. On alternate sides of the paper, score tabs according to general instructions. Fold tabs similarly, in alternate directions.
4. Insert tabs of B all the way into slots of A so that the calibrated scale half is at the "South" mark of A. See illustration 1.

5. Fold tabs of B against the bottom of A in alternate directions. Glue them for a permanent bond, or tape them temporarily if you wish to disassemble the model for mailing or traveling. See illustration 2.

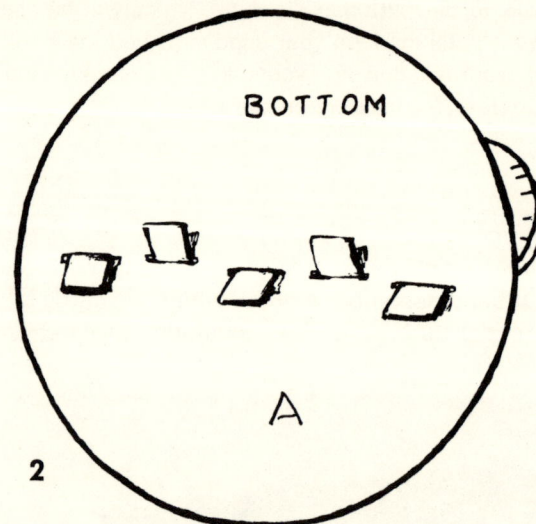

6. Fold up half-circle arm of A so that it slides over the top edge of the circumference of B, as shown in illustration 3.

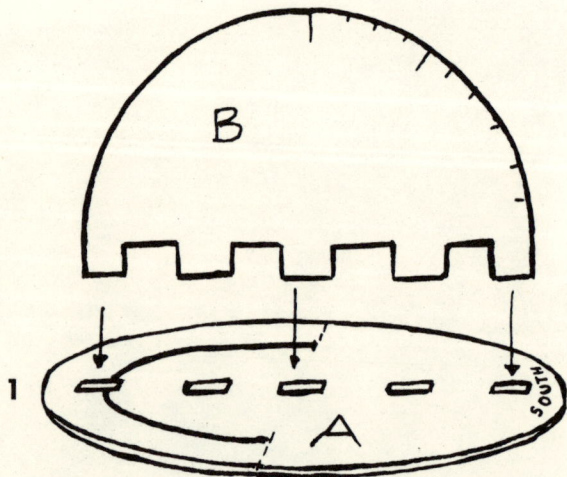

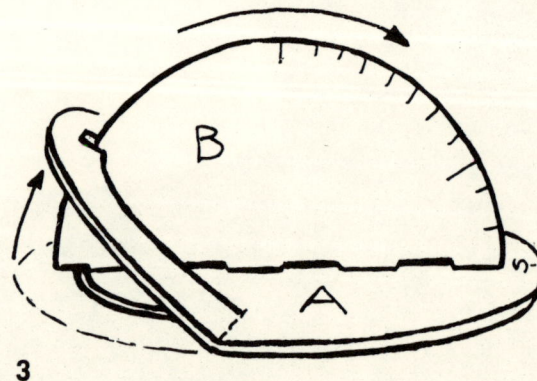

DECLINATIONS OF THE SUN

	Jan.	Feb.	Mar.	Apr.	May	June	July	Aug.	Sept.	Oct.	Nov.	Dec.
1	−23°	−17°	−7°	4°	14°	21°	23°	18°	8°	−2°	−14°	−21°
2	−22	−17	−7	4	15	22	23	17	8	−3	−14	−21
3	−22	−16	−7	5	15	22	23	17	7	−3	−14	−22
4	−22	−16	−6	5	15	22	22	17	7	−4	−15	−22
5	−22	−16	−6	5	16	22	22	17	7	−4	−15	−22
6	−22	−15	−5	6	16	22	22	16	6	−4	−15	−22
7	−22	−15	−5	6	16	22	22	16	6	−5	−16	−22
8	−22	−15	−5	6	16	22	22	16	5	−5	−16	−22
9	−22	−14	−4	7	17	22	22	16	5	−6	−16	−22
10	−22	−14	−4	7	17	22	22	15	5	−6	−16	−22
11	−21	−14	−3	8	17	23	22	15	4	−6	−17	−22
12	−21	−13	−3	8	17	23	22	15	4	−7	−17	−23
13	−21	−13	−3	8	18	23	21	14	4	−7	−17	−23
14	−21	−13	−2	9	18	23	21	14	3	−7	−18	−23
15	−21	−12	−2	9	18	23	21	14	3	−8	−18	−23
16	−21	−12	−1	9	18	23	21	13	2	−8	−18	−23
17	−20	−12	−1	10	19	23	21	13	2	−9	−18	−23
18	−20	−11	−1	10	19	23	21	13	2	−9	−19	−23
19	−20	−11	0	10	19	23	20	12	1	−9	−19	−23
20	−20	−11	0	11	19	23	20	12	1	−10	−19	−23
21	−20	−10	0	11	20	23	20	12	0	−10	−19	−23
22	−19	−10	0	12	20	23	20	11	0	−10	−20	−23
23	−19	−10	0	12	20	23	20	11	0	−11	−20	−23
24	−19	−9	1	12	20	23	20	11	0	−11	−20	−23
25	−19	−9	1	13	20	23	19	10	0	−11	−20	−23
26	−18	−9	1	13	21	23	19	10	−1	−12	−20	−23
27	−18	−8	2	13	21	23	19	10	−1	−12	−21	−23
28	−18	−8	2	13	21	23	19	9	−1	−12	−21	−23
29	−18	−8	3	14	21	23	18	9	−2	−13	−21	−23
30	−17		3	14	21	23	18	9	−2	−13	−21	−23
31	−17		3		21		18	8		−13		−23

HOW TO USE YOUR LATITUDE FINDER

1. Place assembled model on a *level* horizontal surface so that the "South" mark on center line of A (and the edge of the upright B) points a tiny bit to the left of true south, so the shadow will fall on the side of the paper with the printed numbers.
2. At local apparent *noon* (or when the sun is at its highest point, or zenith, of the day) swing the arm of A across the top edge of B until its cast shadow diminishes in width from the broadest to the thinnest possible shadow line, which falls down the side of B to the center point. See illustration 4.

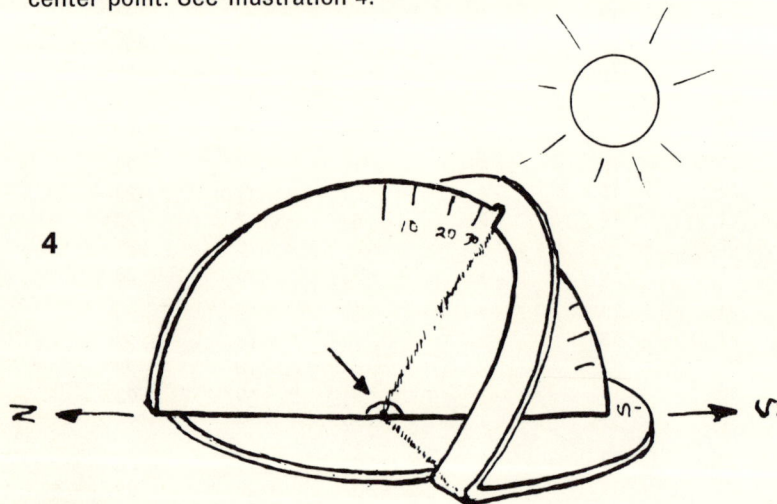

3. Read the number on the calibrated scale where the shadow is cast. Write it down.
4. Read the declination chart (p. 31) showing the sun's declination for your date.
5. Take this number on the chart and add or subtract it (according to the plus or minus sign on the chart for that day) to the number you found in shadow on the model. That number is your latitude. If on March 1, for example, the shadow line reads 50, subtract the −7° on the chart for that date to get your latitude of 43°.

Note: Remember that this number is approximate. Your precision in handling the paper so there is no warping or bending, and the horizontal levelness of the surface you set this dial on, affects its accuracy.

Go ahead and construct the remaining sundials, shifting and turning them in the sunlight to see how the apparent motion of the sun works its own magic apart from the strict regularity of clock time. Afterward, you may want to make the technical adjustments we suggest on pages 42–44 to correct the variation between clock and dial.

Equatorial Dial Variations

Cut out and assemble plates 4, 5, and 6.

You may have seen the next model in other forms but didn't recognize it. The operating principle of the equatorial dial is easy to adapt to many other variations once it is understood. These original versions by Robert Adzema are examples. Instead of using a solid gnomon to cast a shadow, the pierced openings let sunshine project on a dial plate or the ground itself to tell the time, and thus avoids the winter viewing problem of the equatorial wheel. For example, on plate 4, the first digital model, we see the digits travel across the dial table, climb up onto the center strip (where we read the time), cross over it, and continue off to the other side. The hour number appearing on that center strip marks the present hour when the dial is correctly aligned.

DIGITAL SUNDIAL CARD

GENERAL INSTRUCTIONS

1. Remove plate 4 from the back of the book before cutting out parts.
2. Cut only on gray lines.
3. Score all dashed lines (___) on the printed, or top, side of the page. Be careful not to cut through the paper. Score lines with either a pattern-tracing wheel (also called a ponce wheel) or with a dull, slightly rounded tool, such as a butter knife (without teeth).
4. Fold scored lines back away from the side of the paper containing the score.
5. Use a straightedge as a guide for cutting (with an X-Acto blade or mat knife), scoring, and folding all straight lines for the most precise results.

Model 4. "Digital Sundial Card." *Photo by Rosmarie Hausherr.*

SOUTH

© ROBERT ADZEMA 1978

ASSEMBLY

1. Score and fold lines according to general instructions.
   a. Score center dashed line of card A.
   b. Score *solid* line in strip C corresponding to your latitude number from 0 and 0' points outside the strip to across the curved ends of C, as shown in illustration 1. Fold on scored lines.

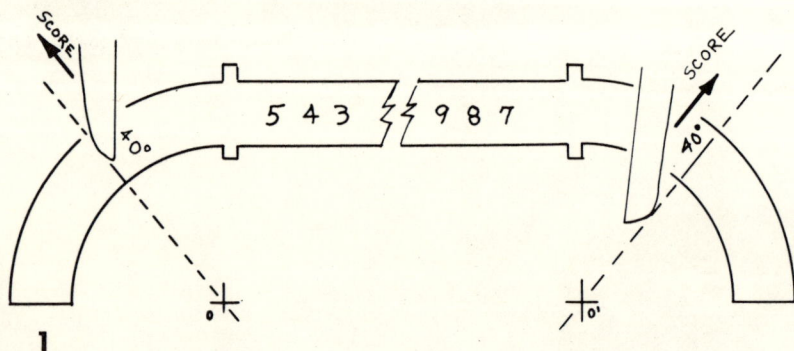

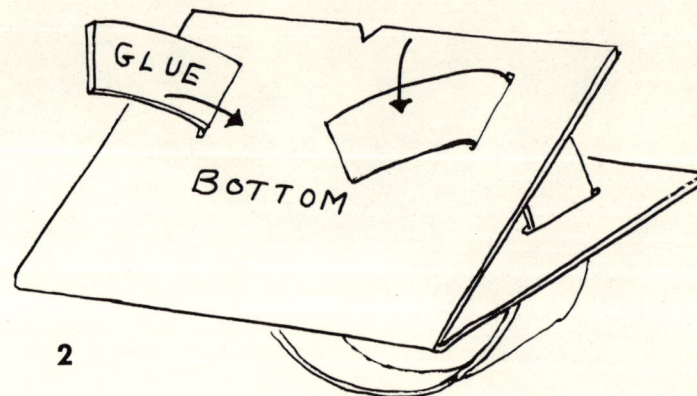

2

c. Open A until the half with strip B reaches stop tabs of C. See illustration 3.

2. Cut out parts.
   a. Cut out card A, including tiny triangular notch indicating north (needed to align dial later) and the four slots.
   b. Cut out strip B and glue it to A where indicated.
   c. Cut out strip C. Cut out numbers *first* with a very sharp blade. Then cut out strip completely. (The numbers are easier to cut before whole part is removed because the small detached strip will move around as you try to cut them.)
3. Assemble parts together.
   a. With A folded in half, so slits align, insert the ends of C from the side of A that B is glued to.
   b. Bend ends of C down and glue them to bottom of A. (If you wish to keep this dial portable for carrying or mailing flat, use detachable masking tape instead of glue for a temporary bond.) See illustration 2.

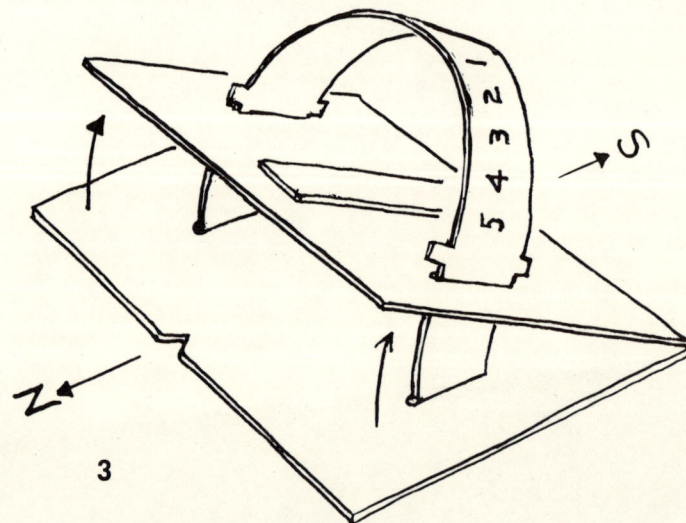

3

HOW TO USE YOUR DIGITAL SUNDIAL CARD

Place dial so "South" mark and north notch align with north–south meridian line and the correct hour appears on B. Read present hour on center strip. Throughout the day the digits travel across the dial plane, climb up onto the strip (where we see the time), cross over it, and continue off to the other side.

You may notice that your assembled sundials tell only the approximate time, not the precise minute. If you align your dial so that the twelve o'clock mark is exactly on the north–south line and check your watch at twelve o'clock Standard Time, you can verify this. Odds are that the sun will cross your noon mark a bit sooner or later than clock noon (unless you are reading the dial on one of the four days of the year when both clock and dial are the same, a phenomenon you will read about in the next chapter).

SINGLE-DIGIT SUNDIAL

GENERAL INSTRUCTIONS

1. Remove plate 5 from the back of the book before cutting out parts.
2. Cut only on gray lines.
3. Score all dotted lines (...) on the blank side, or underside, of the page. On the printed, or top, side make tiny pinholes in last dots of score lines as guide points for scoring when you turn the page over to the blank side. Score lines with either a pattern-tracing wheel (also called a ponce wheel) or with a dull, slightly rounded tool, such as a butter knife (without teeth).
4. Fold all scored lines back away from the side of the paper containing the score.
5. Use a straightedge as a guide for cutting (with an X-Acto blade or mat knife), scoring, and folding all straight lines for the most precise results.

ASSEMBLY

1. Locate your latitude number and its corresponding line on the lower half of the dial. Cut on your own latitude line, following it up into the arch and down to the other side, as shown in illustration 1.

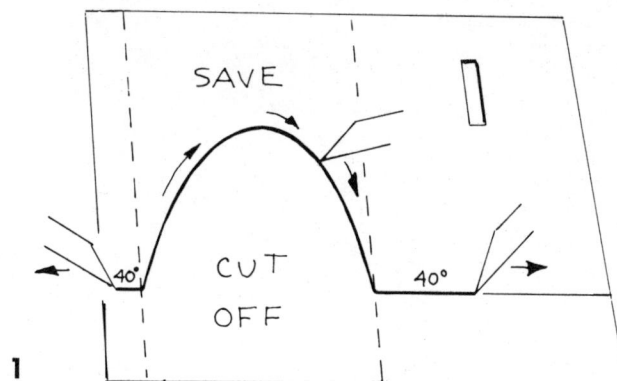

1

2. Cut out A entirely, including slot marked "REMOVE" and the hour numbers. (Use a fresh knife blade to cut out numbers to be sure that cuts are as clean as possible.)
3. On the unprinted, or blank, side of the dial, put a small pencil mark at the bottom edge to correspond to the "South Line" on the printed side of the page. (This mark will be on the outside of your dial after assembly and will be used to align it.) See illustration 2.

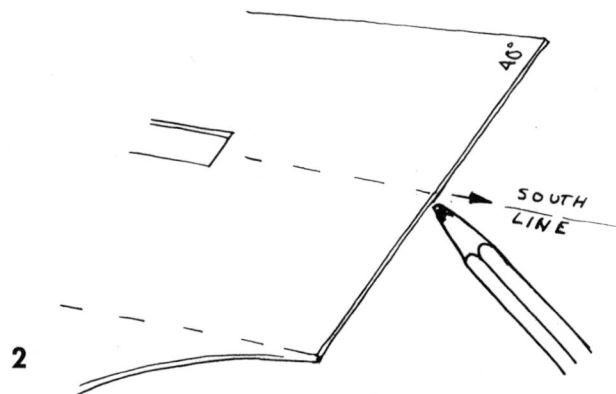

2

35

Model 5. Single-digit sundial (equatorial dial variation).
Photo by Rosmarie Hausherr.

4. On the unprinted, or blank, side of the dial, score both dotted lines according to general instructions and fold tab A and side A in toward the center "N" mark.
5. Curve the paper over to bring tab A to meet side A so that the latitude lines will be on the inside of the assembled dial. Glue the dial closed as shown in illustration 3.

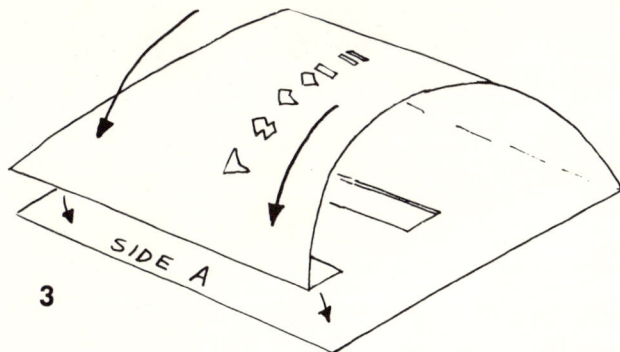

3

HOW TO USE YOUR DIAL

1. Position the dial so that the penciled south mark and printed "N" mark align with the true north–south meridian line. See illustration 4 (which is an overhead view).

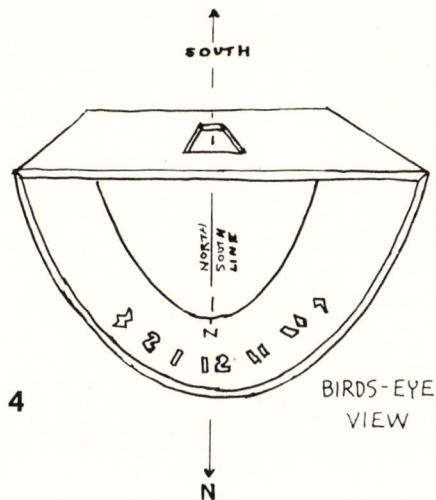

4   BIRDS-EYE VIEW

2. The hour will be projected onto the surface *behind* the dial—one digit at a time.

   Note: During winter months you will see another rectangle of light projected from the dial near your hour number. The reason for this is that the elevation of the sun is so low at this time of year that part of the light spills out of the slot and is projected by itself.

"THE CRACK OF DAWN" (AN EQUATORIAL DIAL)

GENERAL INSTRUCTIONS

1. Remove plate 6 from the back of the book before cutting out parts.
2. Cut only on gray lines.
3. Score all dashed lines (_ _ _) on the printed, or top, side of the page. Score all dotted lines (. . .) on the blank side, or underside, of the page. Be careful not to cut through the paper. Score lines with either a pattern-tracing wheel (also called a ponce wheel) or with a dull, slightly rounded tool, such as a butter knife (without teeth).
4. Fold all scored lines back away from the side of the paper containing the score.
5. Use a straightedge as a guide for cutting (with an X-Acto blade or mat knife), scoring, and folding all straight lines for the most precise results.

ASSEMBLY

1. Cut off the upper left and upper right corners at the number line of your latitude. (Be careful to cut the correct line.) See illustration 1.

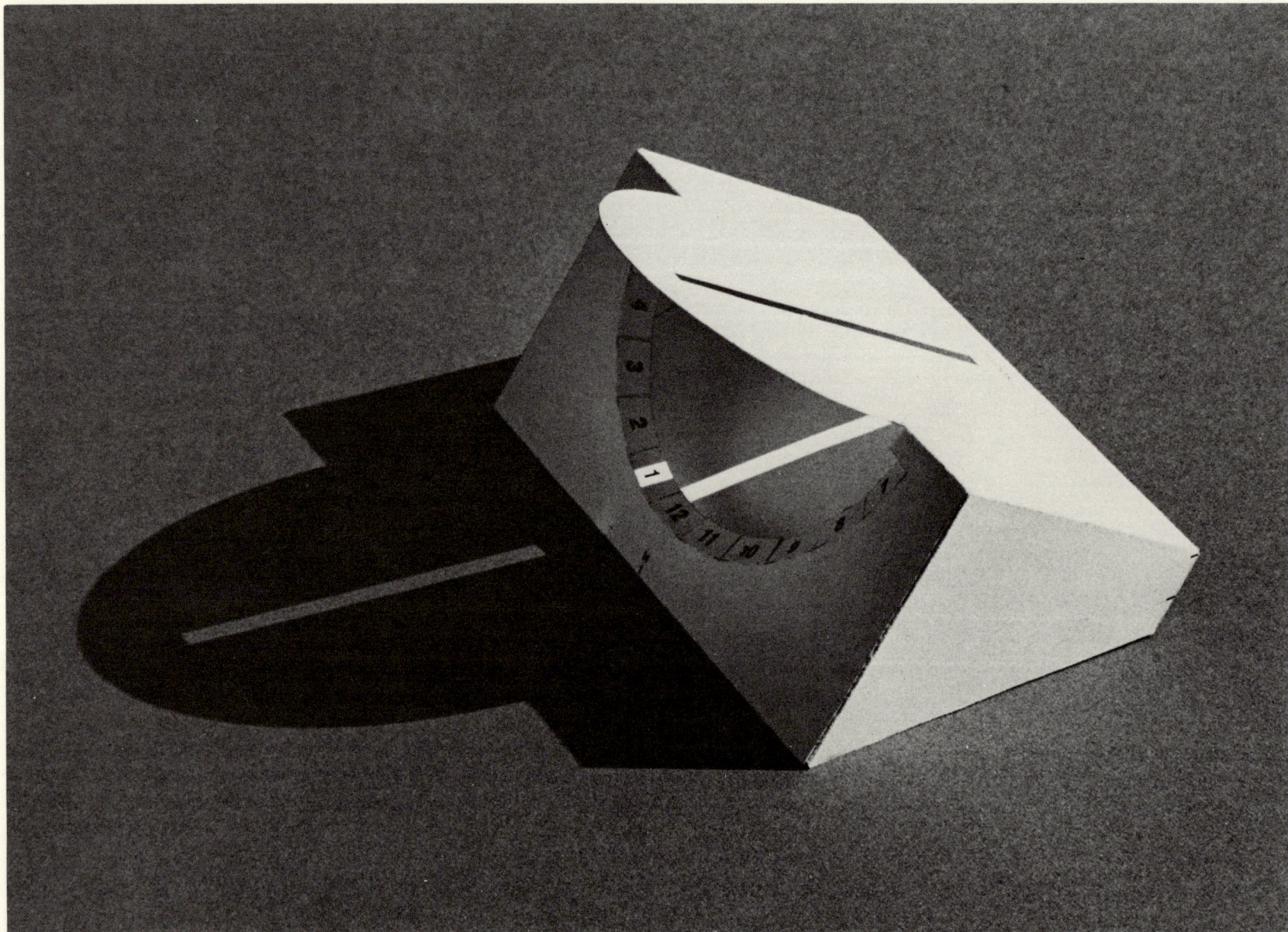

Model 6. "The Crack of Dawn" (equatorial dial variation), view A. *Photo by Rosmarie Hausherr.*

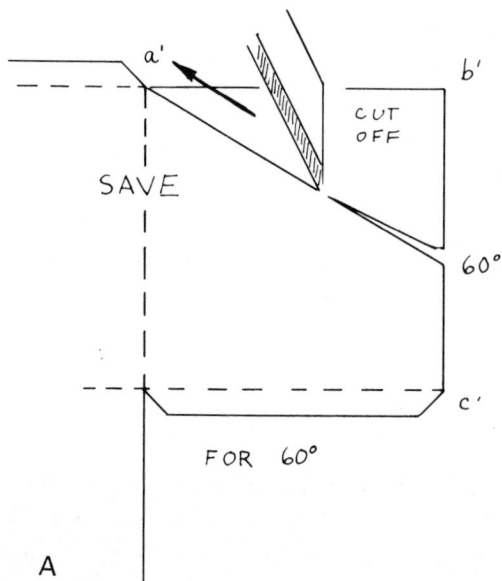

SAVE

CUT OFF

SAVE

a' b' 60° c'

FOR 60°

A

1

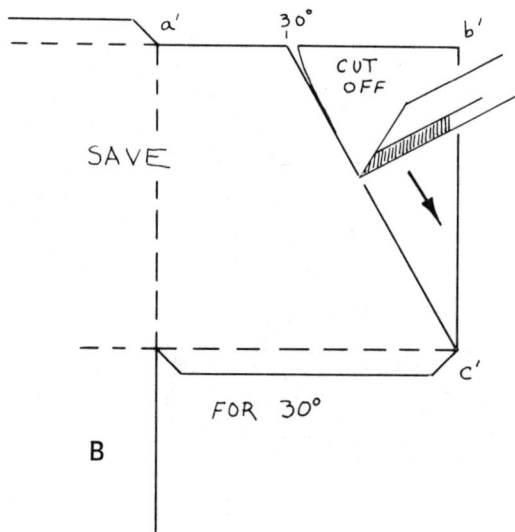

a' 30° b'

CUT OFF

SAVE

c'

FOR 30°

B

a. If your latitude is 45° or more, cut on line b–c (and b'–c') from that number to corner a (and a'). '

b. If your latitude is less than 45°, cut on line a–b (and a'–b') from that number to corner c (and c').

2. Cut out entire dial, including center slot marked "REMOVE."

3. Score all dashed and dotted lines according to general instructions.

4. Fold all scored lines, *but leave the number tabs until later.*

a. Fold side A over to tab A and glue, as shown in illustration 2.

b. Fold side ends up and glue tabs C and C' inside dial, as shown in illustration 2.

c. Fold tabs B and B' on dial bottom to the sides (where you cut off latitude lines) and glue them. If a corner of these tabs extends below the sides, trim it before gluing. See illustration 3.

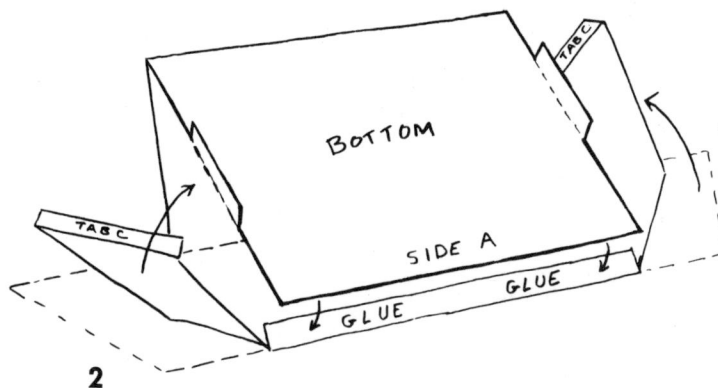

BOTTOM

TAB C

TAB C

SIDE A

GLUE

GLUE

GLUE

2

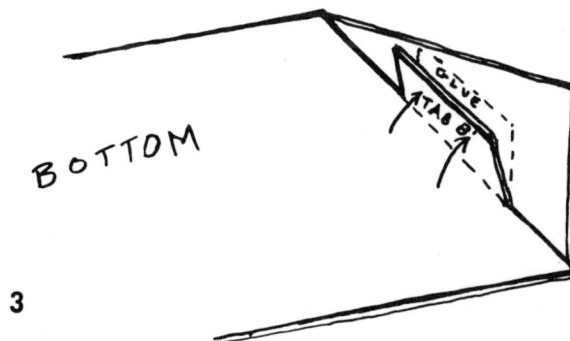

BOTTOM

GLUE

TAB B

3

5. Turn dial upright and press down number tabs at an angle to catch light projected through the slot, as shown in illustration 4 and photograph.

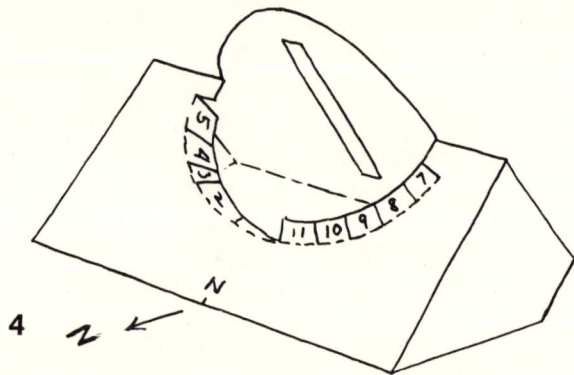

HOW TO USE YOUR DIAL

1. Align dial so that the "South" and "N" marks line up with the true north–south meridian line.
2. Read the hour on numbered tabs as the sunlight illuminates each one.

Model 6. "The Crack of Dawn" (equatorial dial variation), view B. *Photo by Rosmarie Hausherr.*

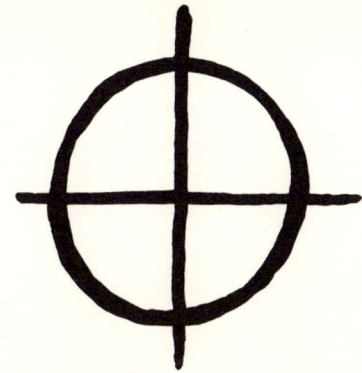

IV
What Sundials Really Do!

Sundials measure *apparent solar time.* The apparent sun is that solar orb we see every day, which appears to revolve around the earth. The *geocentric* view of the universe held by cultures in the West and East alike—from the Egyptians, Greeks, and Romans to the Chinese—until after the sixteenth century originated from man's practical use of this apparent daily path as a guideline. The geocentric theory holds that the earth is the center of the solar system and all other celestial bodies revolve around it. This visual journey of the sun was such a powerful sight that increasingly elaborate explanations had to be created for other observed phenomena that did not fit the system—that is, the observed motions of the other planets. Various astronomers from the third century B.C. until the Renaissance developed theories of epicycles, or little circular orbits within larger ones. Entire philosophical and religious systems based their doctrine around the geocentric model, and it was heretical to think differently. For example, although the Greek astronomer Aristarchus determined that the sun was the center (the *heliocentric* view) in the third century B.C., it wasn't until 1543 that Copernicus, a Polish astronomer, again put forth this idea. When Galileo supported the heliocentric view of the universe in sixteenth-century Italy, he was arrested and forced to recant the idea.

The solar day is the length of time it takes this apparent sun to orbit around the earth to a given meridian line on the earth. It appears to pass over New York City, Chicago, San Francisco,

and so on as our planet turns on its axis from west to east. The *tropical year* is the name given to the amount of time it takes this apparent sun (which, you may have guessed, is actually our earth) to make a complete orbit around the real sun and back to its starting point (approximately 365.24222 days).

However, the amount of time it takes to return to its starting point keeps slowly changing. This is because the earth's axis through the north and south poles is tipped about twenty-three degrees from being perpendicular to its orbit around the sun and because the axis itself is moving in a smaller, slower rotational path, which causes the earth to wobble a tiny bit as it goes around yearly. Another reason why solar time is not consistently equal on a daily basis is that the earth's movement at different rates of speed throughout the year causes our sundials to appear to be sometimes fast and sometimes slow. The Chinese, who used an equal-hour time system, took note of this phenomenon, and by the seventh century A.D. they had worked out mathematical methods of compensating for it.

Explanation: Kepler's Law

The seasonal variation in orbital speed was described by astronomer Johannes Kepler (1571–1630). He told us that each planet travels in an elliptical orbit with the sun positioned at one

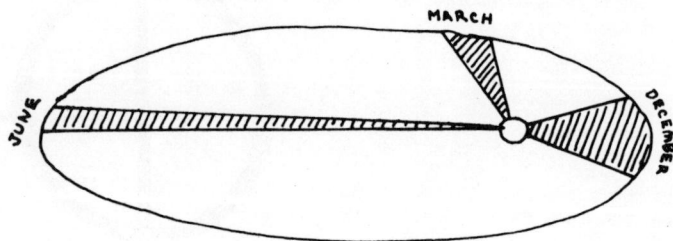

Ellipse describing Kepler's Law.

focus of the ellipse. Our sun is not in the dead center of the path but closer to one of its ends. Kepler's second law stated: The line joining sun and planet sweeps out equal areas in equal times. The areas in the shaded parts of the ellipse shown here (which, for the purpose of illustration, is exaggeratedly sketched out) are all equal. However, the distances along the orbital path in March, December, and June are not. In order for the earth to cover these distances in equal times, it must speed up when it is nearest the sun in this ellipse (about 6 percent faster in January than in June).

Our clocks, on the other hand, run at a constant speed. Therefore, when our apparent sun (that is, the moving earth) travels faster in the winter, clock time falls behind it. Then the mechanical clock can't keep pace with the sun. Think of a race between a steadily moving tortoise (the clock) and a variably paced running hare (the sun). The Ute Indians of North America actually did use the hare as a sun figure. They believed that this animal had to be trained to make its daily rounds. The hare sometimes passes the tortoise and at other times falls behind. The two do get together however on April 15, June 15, September 1, and December 25 (with slight variations as we adjust for Leap Year). On those days our sundials come closest to approximating clock time.

Sundial Adjustments to Make Your Models Approximate Clock Time

You must make two adjustments to get your dials to correspond exactly to clock time. The first has to do with our earth's seasonal speeds in its orbit, and the second is associated with our longitude.

EQUATION-OF-TIME ADJUSTMENT

You already know that the sun's pace varies from that of the clock throughout the year, but now you will have to determine just how many minutes fast or slow it is for your date. This difference between the two is called the *equation of time*. This amount of time is listed on a chart found in an almanac. We have provided a simplified version here. For example, on January 1

EQUATION-OF-TIME CHART
(The Minutes the Sun is Faster or Slower than the Clock)

	Date	Variation	Date	Variation
SLOW	Jan. 1	−3 minutes	July 1	−3 minutes
	15	−9	15	−6
	Feb. 1	−13	Aug. 1	−6
	15	−14	15	−4
	Mar. 1	−13	Sept. 1	0
	15	−9	15	+5
	April 1	−4		
	15	0		
FAST	May 1	+3	Oct. 1	+10
	15	+4	15	+14
	June 1	+2	Nov. 1	+16
	15	0	15	+15
			Dec. 1	+11
			15	+5

Note: Numbers are rounded off to the nearest minute.

the sun (and therefore also your dial) is slower than the clock by 3 minutes, and on November 1 it is faster than the clock by 16 minutes. So at noon on January 1 when you see your dial shows 11:57 instead of twelve o'clock, you add three minutes. On November 1 when your dial says 12:16 at noon you subtract those sixteen minutes of solar time to make the dial agree with the clock hour. If the number of minutes found on the chart for your date exactly corresponds to the difference in time (that you see on your model) between your dial and your clock, you need make *no* further adjustments or calculations on the sundial. Just record these equation of time figures on the dial itself or keep them written separately on a card for easy reference.

If your clock and sundial do *not* agree after you have added or subtracted the number indicated in the equation of time, you must now make a second, longitudinal adjustment. This means that you don't live right on a Standard Time zone meridian line but are east or west of it. Because sundials measure local apparent time, the sun is reaching your place either before or after it goes over the city that is used to calculate the Standard Time for your area. Your dial will be four minutes ahead of your clock for every degree of longitude you are east of that meridian line or four minutes behind for each degree west.

THE STANDARD TIME SYSTEM

The noon sun passes over Boston before it hits New York City, and then goes on to Philadelphia several minutes after that. In other words, when a dial in New York City reads noon, one in Philadelphia reads 11:56 A.M. But all these cities have the same clock time. The Standard Time you hear on the radio is counted from a time zone line, which is more or less in the center of your time zone. For example, the Eastern Standard Time zone meridian runs along the 75° longitude line, which passes through the Philadelphia area. Thus, Philadelphia's local time is the Standard

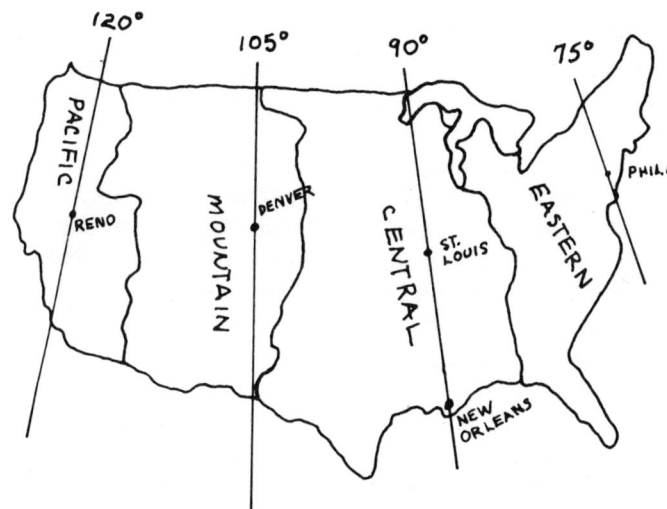

U.S. Standard Time zones. As you can see, the time meridians do not always fall in the center of their respective zones. The shape and sizes of the zones have been adjusted to accommodate political, territorial, and natural land formations. In the Pacific Ocean some time zones even wind around certain islands. It would cause great inconvenience for large cities to fall into two different time zones.

Time for all the cities in the Eastern Standard Time zone. If we did not adhere to this convenient, although nonsolar schedule, people in each place 7½° (or more) east or west of this line would have to set their watches minutes apart to different local times. Only places that are on the same longitude meridian line have the same *local mean time*.

Standard Time is based on a fictitious *mean* sun, which is moving at a constant speed over the earth in contrast to the variable speed of the apparent sun. By the end of a year both the

43

apparent and mean sun finish the race together, unlike the fable of the tortoise and the hare.

Clock time, also called *mean time*, varies according to which time zone you live in. These zones are determined by the distance you are east or west of Greenwich, England. This system was established in the nineteenth century, when it was decided that it would be useful for science and commerce to have a single worldwide time scheme. An international conference drew up the zones (which for political and economic reasons do not always exactly fall on the assigned meridians) and named this system Greenwich Mean Time (also called Universal Time). All twenty-four time zones are calculated from the meridian numbered 0° at Greenwich, with one hour of difference for every 15° of longitude.

So if you want your sundial to compare with an accurate watch set to Standard Time for your zone, you must add or subtract the difference in degrees of longitude of your town and that of the time zone center (plus any Daylight Savings Time adjustments).

Longitude Time Correction Procedure

1. Find out the number of the longitude meridian governing your time zone from any of the sources you used to get your latitude (for example, by map or local agency).

2. Find the difference in the number of degrees between these two locations—the longitude number of your town and the longitude number of the place from which your Standard Time is given.

3. For every degree of longitude difference you are east of that meridian, *subtract* four minutes from your sundial time (or *add* four minutes for each degree if you are west of it).

4. You can incorporate this correction into your previously aligned equatorial-type dials by turning the dial (keeping the "SOUTH" mark stationery as a pivot and pointing south) and turning the other end so that you advance or retard the shadow 1° of a circle for each four minutes of time difference. If you are east of the meridian your dial is fast, so push the north end of the dial toward the east to set back the shadow. If you are west of the meridian your dial is slow and you move the north end toward the west to push it ahead. (In the case of the wheel dial, keep the whole gnomon aligned on the north–south line and merely rotate the round disc to move the shadow.)

V
The Armillary Sphere (Celestial Bracelets)

What is eternal is circular and what is circular is eternal. . . . Unceasing motion is motion in a circle; and this is plain not in theory but in fact. Therefore the first heaven (the circular outer sphere of the universe where the fixed stars are) must be eternal.
Aristotle, *Metaphysics*

Everything celestial was thought to be circular and eternal. From Paleolithic times on through the ages, constant obsession with the symbolism of the circle, not only as a symbol of the sun but also as a philosophic metaphor of all life, shaped all concepts of the universe. Our model here of the armillary sphere is a skeleton map of the cosmos as the ancients saw it, a maquette of an early astronomical instrument used for stellar observation, and also a version of the equatorial sundial.

Cut out and assemble plates 7 and 8.

ARMILLARY SPHERE

GENERAL INSTRUCTIONS

1. Remove plates 7 and 8 from the back of the book before cutting out parts.
2. Cut only on gray lines.
3. Score all dashed lines (_ _ _) on the printed, or top, side of the page. Score all dotted lines (. . .) on the blank side, or underside, of the page. Be careful not to cut through the paper. Score lines with either a pattern-tracing wheel (also called a ponce wheel) or with a dull, slightly rounded tool, such as a butter knife (without teeth).
4. Fold all scored lines back away from the side of the paper containing the score.
5. Use a straightedge as a guide for cutting (with an X-Acto blade or mat knife), scoring, and folding all straight lines for the most precise results.
6. Remember, it is easier to cut out tiny slots with an X-Acto blade before cutting the whole piece from the page.

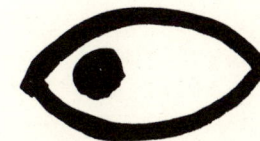

ASSEMBLY

Base (A)
1. Cut out A, and then cut out its large slot wide enough to accommodate two thicknesses of dial paper.
2. Score and fold dashed lines according to general instructions.
3. Fold up side A. Glue tab A inside side A. See illustration 1.
4. Fold up remaining sides, gluing tabs inside them. See illustration 1. Hold until glue is dry.

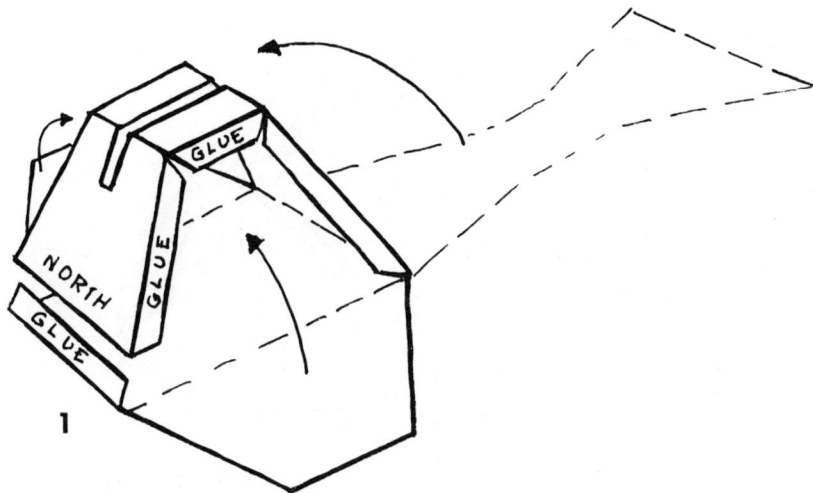

1

Ecliptic–Zodiac Ring (B)

1. Make all interior cuts before cutting out whole piece.
   a. On inner earth support ring, cut out slots E, x', and y'.
   b. On zodiac ring, cut out "JUNE" slot and "DEC" slot.
2. On the printed, or top, side of the paper, near "MARCH," score the dashed lines of the connecting hinge on tab d^2.
3. On the blank side, or underside, of the paper, near "SEPT," score the dotted lines of the connecting hinge on tab d^1. (Poke a tiny pinhole as a guide point through the last score dot at end of each line. Then when you turn over the page to the blank side, you can connect these dots with your score line.)
4. Start first by separating the inner earth support ring from the outer zodiac ring. Cut out the whole double ring B. Be careful when you cut near the hinges so as not to cut through them. See illustration 2.

Latitude Ring (C and F)

1. If you are using an X-Acto blade or mat knife, on the interior edge of ring C, cut out the four slots. Then cut out the whole piece. If you are using scissors, reverse this cutting order.
2. Cut out latitude support (F). Glue it to the blank lower half of C (so that you have numbers on both sides of the ring).

Model 7–8. Armillary sphere with base. *Photo by Rosmarie Hausherr.*

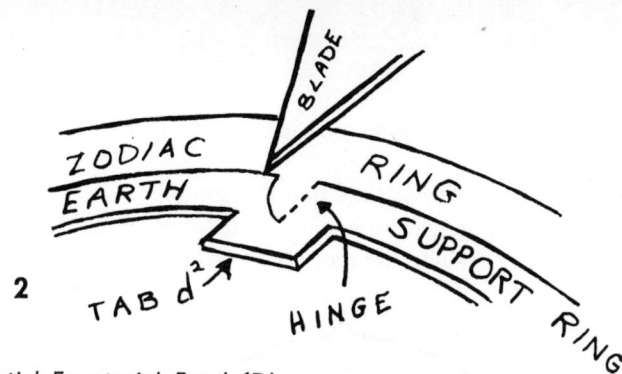

2

Celestial Equatorial Band (D)

1. Cut out slots in D at "6 AM," "NOON," "6 PM," and at both ends.
2. Cut out D completely. (Again, reverse this cutting order if you are using scissors instead of an X-Acto blade.)

ASSEMBLE ALL PARTS

1. Cut out earth (E) and insert it into the center slot of inner earth support ring (which is hinged within ecliptic zodiac ring B). Glue by placing glue along slot.
2. Inside ecliptic ring B, loop the equatorial band D around the center gnomon (carrying E) so that the numbers face inward. Glue it so the outer slots of D overlap. See illustration 3.

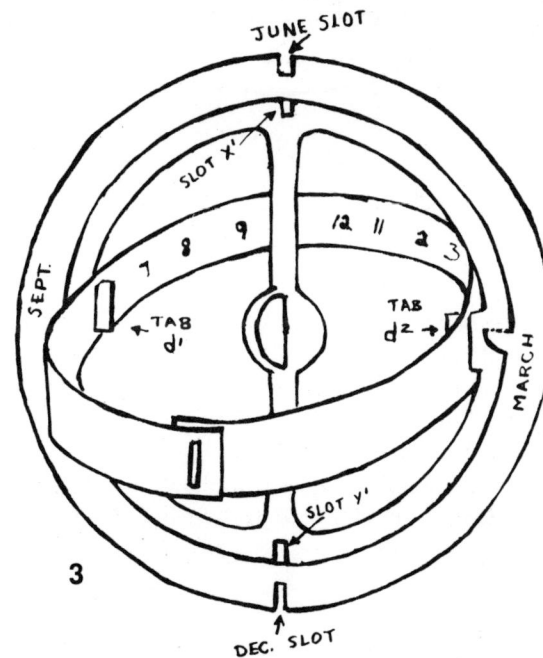

3

3. Place band D on B, inserting tabs d¹ and d² into "6 AM" and "6 PM" slots. See illustration 3.
4. Join B and C at intersecting slots. See illustration 4.
   a. "JUNE" slot of B should intersect "JUNE" slot of C.
   b. "DEC" slot of B should intersect "DEC" slot of C.

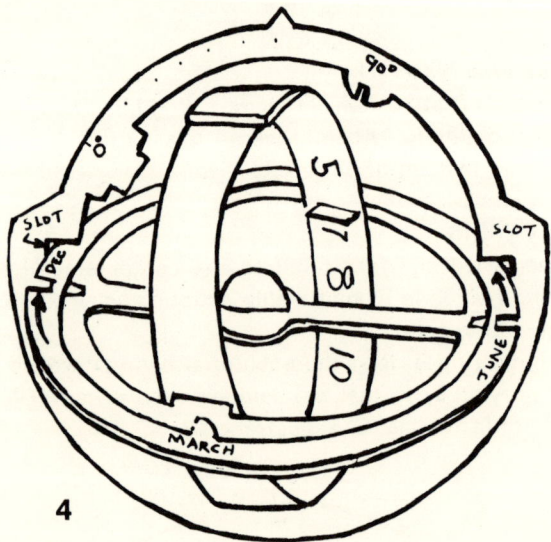

4

5. Join remaining slots in D to tabs of C by tilting earth support ring. See illustration 5.
   a. Tab d³ should fit into overlapped end slot of D.
   b. Tab d⁴ should fit into "NOON" slot of D.
6. Put glue on all joints and let dry.
7. Join base A and assembled sphere by inserting the bottom (double thickness) of C into the slot of A so that the 90° mark faces south. *Do not glue.*

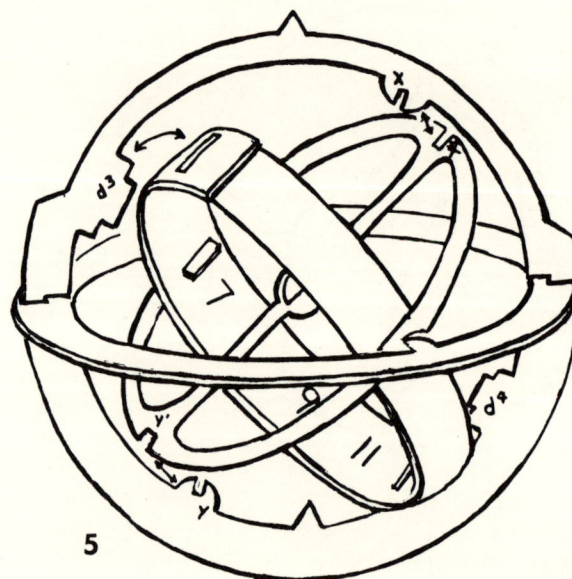

5

HOW TO USE YOUR ARMILLARY SPHERE AS A SUNDIAL

1. Rotate the entire sphere in the base slot until your latitude number lines up with the "V" marks on top of base. See illustration 6. (Now you may glue your sphere to your base if you wish to make the joint permanent for only your latitude.)
2. Align the "NORTH" and "SOUTH" marks on your base with the true north–south meridian line.
3. Read the time at the shadow cast by the central gnomon onto the hour marks of band D.

The rings, or *armillae* (Latin for "bracelets"), show the major motions and spatial coordinates of the sky. The small three-dimensional globe at the center on the gnomon bar represents the earth according to the geocentric view of the solar system.

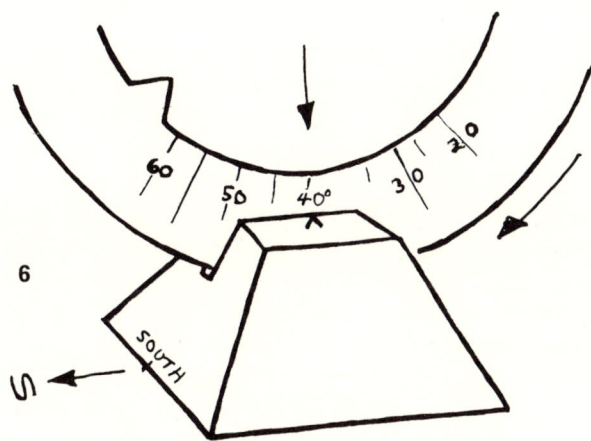

Note: You may also hang the armillary (preferably separated from the base) by piercing a small hole in the upper part of the latitude ring (C) and suspending it by a piece of thread. Remember, however, it will not tell time unless you tilt it to the proper angle for your latitude number.

When you position this model so that the center-bar gnomon is parallel to the north–south meridian line, you also have an equatorial sundial with the gnomon shadow falling across the hour marks on the equatorial band. This band is an imaginary extension of the earth's equator out into space. An imaginary center line of this band is called the celestial equator.

Both the Chinese and the Greek astronomer Ptolemy used much larger versions of this sphere as an observation tool around the second century A.D. by installing sighting tubes on a center bar on their movable ecliptic rings, with which they could study the stars. The tubes shielded the eye from extraneous light from other parts of the sky during observation, thus making stars that were faint in the sky more visible to the eye. However, the in-corporation of tubes into armillary spheres only became standard in China around A.D. 725. Western Europe did use these accessories on other navigational instruments, such as the Renaissance cross-staff and the eighteenth-century sextant.

The ancestor of the armillary was not our earth's globe. Ironically, the celestial sphere preceded the terrestrial model of our planet by nearly a thousand years (even though the Greeks knew and proved mathematically that the earth was round). Eratosthenes deduced this fact in 270 B.C. from measuring shadows of gnomons at different locations and applying geometrical calculations to the data. The first recorded earth globe in Europe was the "Apple of Master Behaim" in A.D. 1492 (the year, of course, when Columbus set sail for America).

The origin of the idea for the armillary was the *planisphere*, a sky chart that mapped the heavenly sphere and its star positions on a flat surface. The planisphere developed both in the Near East and in China before the first millenium B.C. and included demarcations for the celestial equator, the ancient polar star, and concentric latitudelike circles to mark spatial coordinates (like the ones found on maps and globes). The hemispherical sundial bowl of the Babylonian astronomer Berossus (about 270 B.C.), which had a vertical gnomon, may have also influenced the development of the armillary sphere.

The first primitive form of the true armillary in China was a single ring with marks on it at intervals corresponding to spaces between the stars around the north celestial pole. These marks served as positioning guides when lined up with the eye in order to observe that area of the sky. It was aligned parallel to the earth's axis, just like the equatorial dials you made in chapter 3. Over successive centuries, the Chinese eventually installed a model earth at the center of their armillary, made solid representations of the celestial globe with star maps of all constellations, and added balls to their open spheres to represent planets. One possible reason they might have continued to use this model after the West developed more advanced astronomical instruments was because the armillary sphere was also a sacred divination tool for predicting the future.

The Armillary as an Instrument of Magic and Power

Astronomical information was the secret power of priest kings in early China, and the astronomical observatories were both cosmological temples and the emperor's ritual home. Priests held onto this power because in their agricultural economy an accurate calendar was needed for planting and harvesting. Those who could read the skies and supply this information naturally became the leaders.

THE RITES OF CHOU

A record of Chinese science and religion called the *Chou Li* ("The Rites of Chou," dating from the second century B.C.) discussed the role of the royal owner of the armillary. The emperor's duties were to (1) fix the four cardinal points of the sky through observations of the sun and polar stars (using the *hun*, or armillary sphere), (2) make diagrams of the state of the heavens, (3) observe the solstices and equinoxes in order to determine the progress of the seasons, and (4) predict the coming of floods (monsoon season) and droughts, abundance and famine. During the day the emperor recorded the lengths of the sundial shadows, and at night he watched the movements of the stars with the armillary sighting tubes. His sacred tool was a movable ring version of your own celestial sphere.

Prophecy of prosperity and misfortune was risky, since the astronomer-emperors were held responsible for influencing events. Solar rulers were also inconvenienced by ritual taboos that governed every aspect and gesture of their daily lives. For example, the Egyptian pharaohs had to walk around the temple every day symbolically to assist the sun's successful journey. The Japanese mikado, who was a male incarnation of the female sun goddess and the dynamic center of the universe, could not make an extraneous bodily move lest he disturb some part of nature.

So these political rulers transferred the burdens of prediction to a special class of imperial astronomers who were punished for their forecasting failures. According to legend, after the Chinese separation of ruler from astronomer, the emperor Yao commissioned four astronomer-magicians to go to the four corners of the earth in order to turn back the sun at each solstice and to keep it moving at each equinox. They were also responsible for the prevention of eclipses. When the quartet failed (naturally), the emperor pronounced the death sentence on each of them.

The medieval Chinese court astronomer was considered a sage who traced connections of events in nature by interpreting positions of celestial bodies. He also prepared calendars and star charts. The owner of the armillary in this age was evidently responsible for more modest and more ambiguous prophecies.

For example, a legendary Chinese Buddhist astronomer (mentioned in a scroll dated A.D. 855) sought mercy for an imprisoned murderer who had helped him in his youth. I-Hsing, the astronomer, went to the *Hun-Thein* (Temple of the Armillary or Celestial Sphere) and ordered his workers there to place a large open pot in one of the temple rooms. Then he sent two servants into a nearby garden with instructions to catch and put into this pot anything which came to the garden in the exact number of seven. At nightfall seven pigs arrived. These were caught by the workers and placed in that covered pot, which was then inscribed with secret words. When the emperor discovered that the seven stars of the Great Bear constellation disappeared that night, he was apprehensive and sent to I-Hsing for an explanation. The astronomer-sage declared that the event was an important omen, a warning that the emperor could prevent catastrophe and influence the stars in favor of life rather than death by a demonstration of his own mercy and compassion. The emperor agreed with this interpretation and pardoned the criminal. Mysteriously, the seven stars of the Great Bear reappeared later in the sky, and when the pot filled with pigs in the Temple of the Armillary was uncovered, it was empty.

ARMILLARIES OF THE WESTERN WORLD

Many ancient Western astronomers and mathematicians used armillaries, including Eudoxus of Cnidus (who made a star map using one in the fourth century B.C.), Hipparchus (who rediscovered the precession of the equinoxes), and Archimedes in the third century B.C., Eratosthenes (who calculated the circumference of the earth in the third century B.C.), and second-century-A.D. astronomer Ptolemy (who called it an *astrolabon organon*, or "star-taking organ," from the Greek *astro* meaning "star," *labano* meaning "to take," and *organon* meaning "organ").

The Arabs called the armillary "the owner of the rings." They used simple versions with sighting tubes by the twelfth century A.D. and added horizon and equator rings in the thirteenth century. The Spanish Moors supplied armillary prototypes for all later European models. This sphere was used not only for astronomical viewing and measurement but also as a demonstrational model of the solar system.

The circle was such a powerful magic image that philosophical and religious obsession with it hindered the progress of astronomy up to the Renaissance. Astronomy, mathematics, philosophy, and religion were all interconnected. The priests of many ages were scholars and astronomers as well, and the temples were sometimes the only centers of learning. The circular celestial orbits with the earth in the center became symbolic of the church's fixed view of man and his place in the universe.

The cumbersome number of rings added onto the armillary in order to illustrate these circular astronomical theories finally made the device unusable for anything but decoration of gentlemen's libraries. Because the craft of precision metalworking was less accomplished then than it is today, the sheer weight of the circles deformed points of juncture at the equinoxes, causing errors in the arcs. However, the Chinese, who never accepted the Western planetary schemes, continued to use their simpler versions, and mechanized them with water power so that they would move with the rotation of the stars around the north polar star in the eleventh century. The West never efficiently mechanized their spheres because of the complexity of parts making up eccentric planetary orbits. The Orient used the armillary as an astronomical sighting tool well into the seventeenth century, even after the West had discarded it in favor of other scientific devices. However, the armillary still exists today in sundial form. This modern example by the late artist Paul Manship combines sculptural figures with the time-keeping function. *The Garden of Eden* eliminates some of the rings and simplifies reading of the dial.

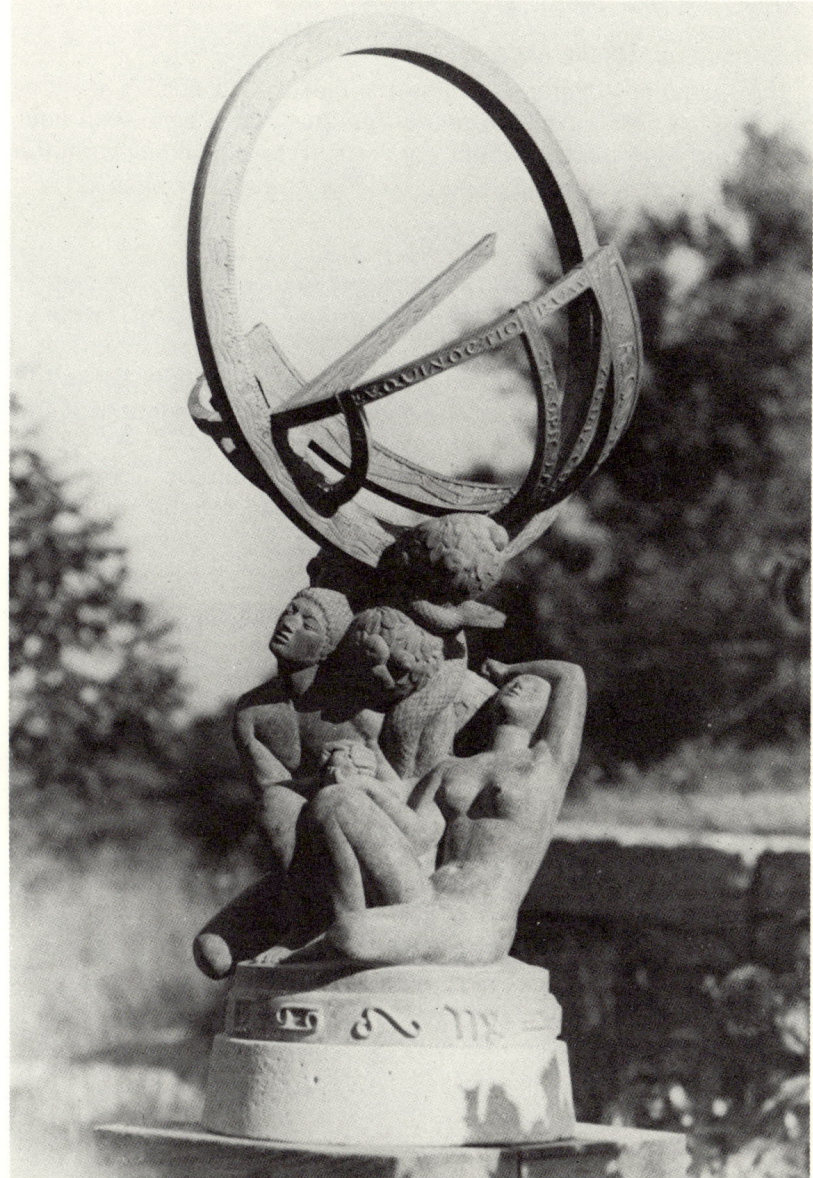

Modern sculptural armillary sphere sundials by artist Paul Manship (1885–1966) join art with gnomonics. *The Garden of Eden. Photo courtesy of John Manship.*

51

Parts of Your Armillary Sphere

CELESTIAL EQUATORIAL BAND

The *celestial equator* is a line parallel to the earth's equator out in space. We have broadened this line into a band with hour marks on your model, which converts it to a sundial with the center-axis gnomon projecting a shadow onto the hour marks.

THE ECLIPTIC-ZODIAC RING

This is the yearly path of the apparent sun, or—in the geocentric world view—the real sun traveling around the earth. The inner earth support ring attached inside the ecliptic-zodiac on your model holds the central earth in place while its middle bar is the sundial gnomon, which casts its shadow on the hour marks of the equatorial band.

In reality the *ecliptic* is elliptical in shape, but we have reproduced circular rings such as those found on armillaries before Kepler's day. This ecliptic circle bounds an imaginary surface called the *plane of the ecliptic*. As you can see on your model, this plane is tilted at an angle so that it rises above and falls below the equatorial band. This seasonal north-and-south motion is due to the tilt of our earth's axis. While the axis itself points toward the North Star, our earth's tilt is such that in winter the northern hemisphere leans away from the sun and in the summer it tips toward the sun. When the apparent sun reaches its northernmost point on this orbit (June 21), we are at the summer solstice (the longest day of the year for the northern hemisphere) and the noon sun reaches its greatest height, or *declination*, away from the equator. At the points where the ecliptic intersects the equatorial band, called *colures*, we have the equinoxes of equal day and night (March 21 and September 23). At the lowest or southernmost juncture, where the two rings are far apart again, is the December 21 winter solstice. Some old armillaries have two additional rings, parallel to the equatorial band, at these northern and southern joints called the Tropic of Cancer and the

Mary Lucier's "Dawn Burn."

A. Video screen showing the start of the first day's dawn sun.

B. Sun's trajectory path as it burns the video tube.

Tropic of Capricorn because the sun used to enter the constellations of Cancer and Capricorn, respectively, at those times of year. The sun's declination changes significantly even from day to day. This can be seen in artist Mary Lucier's "Dawn Burn," a video recording of the rising sun during seven mornings of videotaping. Each sunrise burns a black path on the video camera tube according to the variations in the sun's actual trajectory each morning while the video camera remains stationary throughout the period. These changing positions of the sun north and south, as well as the uneven rate of speed in our yearly path, create the difference between clock time and solar, or sundial, time (which is based on the apparent sun).

Claudius Ptolemy, the most influential astronomer of the ancient world, could only explain the seasons by his theory that the sun moved in a conical spiral down the sky and back up each year. At the winter solstice, the sun was the farthest distance away from the earth. It stopped there momentarily (*solstice* means "the sun had stood still") and then started the climb upward toward the equator and inward again. The earth became warmer as its spiral path narrowed in the summer, bringing the sun and earth closer together, and then became colder again when it expanded at the southern end below the equator in the winter.

Explanation of the Zodiac

You will find the zodiacal symbols on the ecliptic ring. The zodiacal constellations are found in the sky within a narrow strip surrounding the center line of the ecliptic.

Place your armillary sphere on a table in the center of an open space where you can move all the way around it. Let the center represent the sun instead of the earth while you play the part of the moving earth circling the sun. This is the heliocentric view of the solar system. From your sight line as you walk around it,

C. **Sun's appearance on a later day with the previous burn.**

D. **Two solar burn patterns on the video camera tube.** *Photos courtesy of Mary Lucier.*

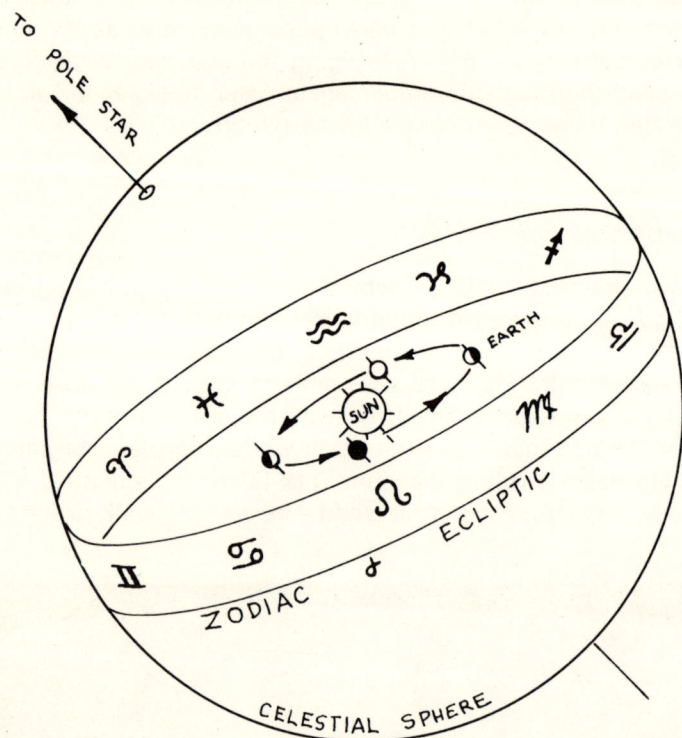

The celestial sphere. The ecliptic or path of the apparent sun runs along the center of the zodiacal band.

rise and juxtaposing their position with the sun's appearance and disappearance each day. The stars near or at that position of the sun are called *heliacal*, from the Greek *helios*, meaning "sun."

The two latitude lines called the Tropics of Cancer and Capricorn mark the declination of the sun at the solstices. However, if you look at the lowest and highest points of the ecliptic ring in those sections marked June and December on your model, you will see that these points (*colures*) do not coincide with the start of Cancer and Capricorn. They fall instead in Gemini and Sagittarius. You may have also noticed that your birthday may not fall into the sign on your armillary which most popular horoscopes tell you it should. In addition, if you look for the heliacal zodiacal constellation outdoors in the sky on your birthday, a different one than you are told to expect may be there. The vernal (spring) equinox actually occurs when the sun rises in front of the constellation of Pisces, but from a popular zodiacal chart you might expect it would be in front of Aries. There are two reasons for this discrepancy:

First, all the constellations on the zodiac are not equal in size, despite the fact that astrologers had neatly divided the group into twelve equal parts by the fourth century B.C. For example, Scorpio is larger in breadth than its alloted thirty-degree arc, and Cancer is smaller. By the way, there are other stars falling within the zodiacal circle, and if they were added, the number of zodiacal constellations would increase to a possible fourteen. Twelve was chosen by the Assyrians and Babylonians because that was the number of full moons within a year (even though in every third year there are thirteen full moons).

the sun appears to progress through a series of constellations on the zodiac. Of course, we on earth see the sun against a backdrop of daytime starless sky, but at night the sky near the ecliptic (where the sun was during the day) is filled with the constellations of the zodiac. They follow each other in sequence across the sky, but each reaches its highest point in the sky a little earlier each night. The astrological expression "The sun is in Aries [or Cancer or whatever]" has traditionally meant that the sun is in front of that constellation. Although we can't see constellations during the daylight, ancient astronomers calculated their position by noting which stars rose and set just after dark and before sun-

The Precession of the Equinoxes

The second reason for the discrepancy between the constellations and the signs on astrological charts is a phenomenon called the *precession of the equinoxes*, which is caused by a small wobble of the earth's axis, which in turn causes the zodiacal constellation to shift in relation to the equinoctial colures.

Look at your armillary model to see where the earth's axis points. This north celestial pole (NCP on your pattern just outside

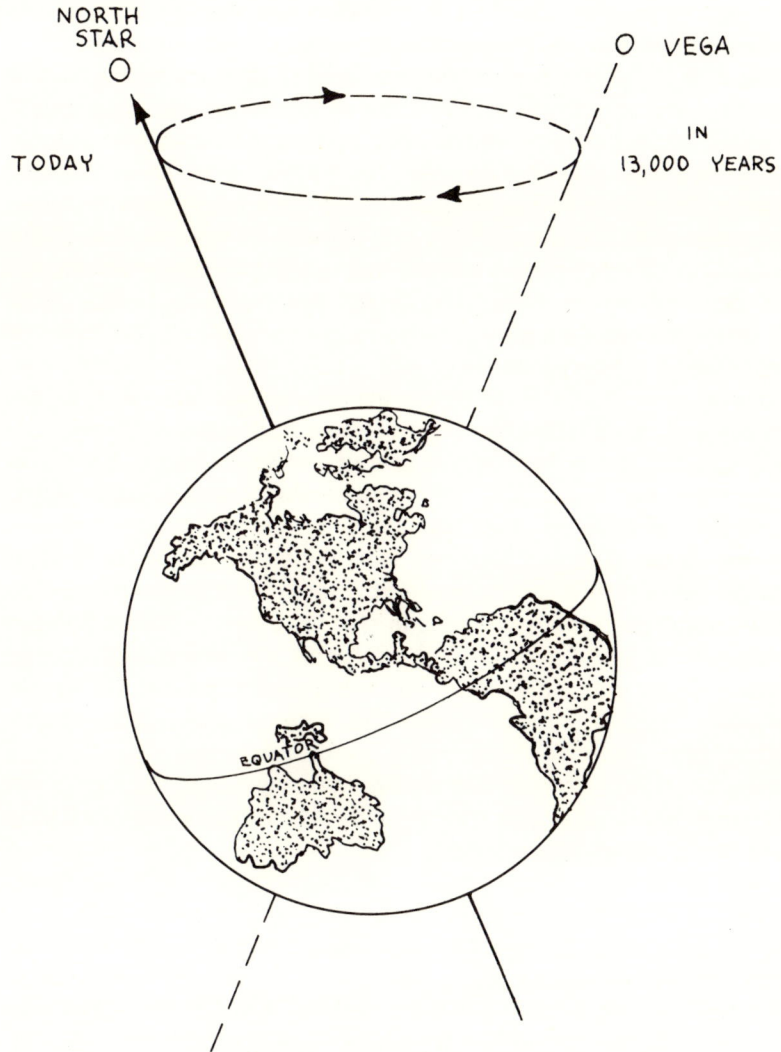

NORTH STAR

TODAY

O VEGA

IN 13,000 YEARS

EQUATOR

The precession of the equinoxes is caused by the slow wobble of the earth's axis so that our pole star will change over thousands of years.

the cutting line at "90°") is an imaginary extension of the axis into space. It now points (both on your model and in the sky) very near to the North Star, or Polaris.

In actuality, however, the earth's axis is not stationary; it moves in a very slow orbit around the north ecliptic pole (NEP—a small triangular projection on your latitude ring C). The north ecliptic pole is directly perpendicular to the ecliptic ring. The north celestrial pole on your armillary is oriented toward where the axis of our earth points right now, although because one revolution of this axis around the ecliptic pole takes 26,000 years, every few thousand years the north celestial pole points to a different star. In 2800 B.C. it pointed to Alpha Draconis, one of the stars in the tail of the constellation of Draco (the Dragon). Now it points toward the tail-end star of the Little Bear, and in the next several thousand years it will move toward another star, Vega, in an entirely different constellation.

The solar, or *tropical*, year is measured by the time it takes the earth to complete an orbit from one point back to that same spot in relation to the sun—in other words, the interval between two successive returns of the apparent sun to a meridian point on the equator. This is measured from the place at which the apparent sun crosses the equator at the spring equinox, and on the celestial sphere of your model it is the colure where the ecliptic and equator intersect in March. However, the wobble of the earth's axis (which is also slowly moving westward) pushing that equinox point on the equator forward each year, plus the variable speed of the earth's journey on the ecliptic, causes a few minutes' variation in the length of the year. This tropical year, whose name is taken from the Latin *tropicus*, meaning "belonging to a turn of the sun," is approximately 365.24 days in length.

Because of yearly solar time differences, astronomers employ the *sidereal year* (from the Latin *sidus*, or "star") for their calculations. This unit remains constant and is determined by the interval of time it takes the earth to orbit completely around to that same spot in relation to another star. The sidereal year is 365.26 of our tropical days (which is equal to 366.26 sidereal days) in length.

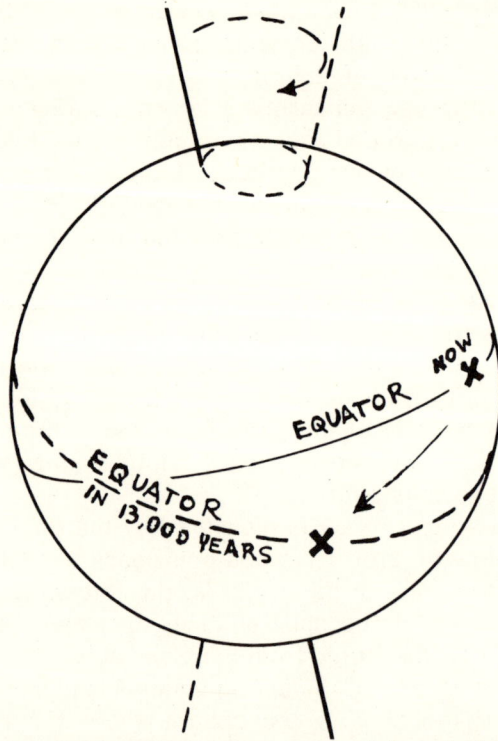

Precession pushes the equinoctial points westward through the ages.

In order to visualize the reasons for the differences between the star year and solar year, imagine that you are the apparent sun. You are on a carousel (that is, the ecliptic), moving in orbit past a stationary platform (a star) where you catch a brass ring (a vernal equinox). You see that if the brass ring station stays in one place, you will get the ring (equinox) at the same location (or place among the stars) as you go around, and you will also catch it at even intervals of time. This imagined ride represents the constantly equal star, or sidereal, year.

This next trip around depicts the tropical, or solar, year. Now you are orbiting on the carousel toward a brass ring station (now

a point on the earth) which is moving (unlike the star) but is also leaning toward you (as its axis wobbles toward you very slowly in the precession of the equinoxes). In addition, the station is tilting (the axis tilts twenty-three degrees), and someone on that moving platform reaches out in order to deliver the ring (earth location) right into your hand. Now you are not only going to get that ring faster (a shorter solar year) but also at a different place in the circle (different point on the ecliptic) each time you go around. The locations of the equinoxes as they occur over the earth meridians are not stationary (as is the case on your model); they really move to slightly different points through the years. This succession of equinox positions is the origin of the term *the precession of the equinoxes.*

This means that the position of the sun on the day of the vernal equinox, marking the start of the spring season (March 21) and signaling the rebirth and return of the agricultural deities for many early societies, will over thousands of years occur retrogressively in front of one zodiacal constellation after another— and our polar star will change also. Pisces became the heliacal constellation rising with our vernal equinox sun around 6 B.C. (near the birth of Jesus). Pisces is still the vernal equinox heliacal constellation, but Aquarius will soon (around the year 2200) replace it. Therefore astrologers say we are currently in the Age of the Fish (Pisces), moving into the Age of the Water Bearer (Aquarius).

When the constellations that marked each age moved in the sky to beneath the celestial equator, they were said to have plunged into the heavenly sea, and myths about the deaths of heroes and world catastrophes (for example, the biblical flood) as the "Golden Age" (roughly from 4000 B.C. through A.D. 60. It was also called the Saturnian era because that period was supposed to have been ruled by the planet Saturn. Saturn became identified with an ancient agricultural deity, Kronos (not Chronos), who later became known as the Greek god, Phaethon.

The Greek myth of the fall of Phaethon is a celestial geography lesson, an example of how folklore served as an educational introduction to astronomy. The tale describes the motions of the apparent sun within the celestial sphere. Phaethon's catastrophe in the sky is one chronicle of the precession of the equinoxes.

THE MYTH OF PHAETHON

When Phaethon, son of Apollo and the nymph Clymene, failed to get the respect from his peers that he felt due to his royal blood, he went to his father, the Sun, to demand he be given proof of his station. After Apollo promised to grant it by any means his son chose, Phaethon demanded to drive the chariot of the Sun across the sky for a day. Apollo tried to dissuade the youth by telling him of all the dangers: the problems of staying on the course with a constantly moving landscape, the frightful monsters—such as Taurus the Bull, the Archer, Leo's jaws, the place where Scorpio the Scorpion reached out its pincers in one direction and Cancer the Crab extended its claws in another, and the like. Finally, since Apollo couldn't discourage Phaethon, he gave him driving instructions: Don't take the straight road between the circles but turn off to the left, and keep within the limit of the middle zone (that is, between the tropics).* Phaethon set out bravely, but the horses felt the lighter weight of the boy in the chariot behind them and took off from the accustomed route. When Phaethon saw the monstrous scorpion, he let the reins fall, and the steeds plunged into unknown regions. The world was scorched and shaken so badly out of place that Jupiter shot down the errant driver Phaethon into Eradanus, the great river that is below the celestial equator.

Phaethon, the planet Saturn, who was believed to rule the Golden Age, was called a wandering star, in contrast to the fixed stars in the constellations. It was believed that his erratic course through

* The path of the ecliptic curves 23° north and south of the celestial equator. The circles may possibly represent what is now called "the circlets" in the constellation Pisces. This point in the sky was of importance. Pisces today is at the equinoctial point, the place where the sun crosses the celestial equator at the spring, or vernal, equinox. But at that time the Fish were north of the equator. This group of stars opens out into a V shape toward Pegasus. If Phaethon had gone straight up the middle of the V instead of turning, he would have ended up in Pegasus and off the route.

the sky caused the shifting of the stars and the scorching of the earth. The ancients found the observed motions of the planets to be strange or erratic because they did not follow the same apparent motions as the rest of the stars. Their orbits never fit well into the circular geocentric systems. The scorching fire was caused by newly rising stars climbing in the sky from the celestial equator. This fire touched the earth until the stars had risen to a sufficient height. By that time, of course, Phaethon had plunged into the sea, and the precession of the equinoxes had shifted the world into a new age.

Another mythical description of precession is this California Chumash Indian story of the Coyote's visit to the sky:

THE COYOTE AND THE SUN

The Coyote visited the Sun one day and begged to be allowed to follow him on his daily journey. The Sun tried to dissuade him but finally gave in to the Coyote's begging. They followed the Sun's usual trail, which was a cord stretched around the world. After a while the Coyote badgered the Sun to let him carry his firebrand. The Sun lectured him on all the dangers of this task and warned him that if he made a mistake, he would burn the world. Eventually the Sun again gave in to the Coyote, who started off all right but then let the firebrand slip from his paws so that it started to fall. The world nearly burned up before the Sun could retrieve it. The next day the Coyote wanted to go home to the earth and hitched a ride on the back of one of two descending eagles. The Coyote became jealous of their ability to fly and started plucking feathers from the back of his eagle's neck. The angry bird threw his passenger off, and the Coyote fell to the ground.

Similar stories abound throughout the world about the precession of the equinoxes, with the actors taking different names in each land. For example, in the Bible Sampson is the one who pulled down the pillars of the system; in Mexico it was the Black and Red Tezcatlipoca; and in Japan Susanowo (whose name means Brave-Swift-Impetuous Male) descended to become king of the underworld after he tore apart his own palace. As these figures disappeared, "new fire" was kindled as new stars took their places on the "true earth" (or celestial sphere).

Today the armillary sphere has been stripped not only of many of its astronomically useful parts but also of its mythological heritage. You may see it in sundial form in public parks and gardens, but it is now a mere skeleton of the original celestial sphere. Sometimes all that remains is a central gnomon pointing to the north celestial pole, a circular band (or even half-circle) to receive the cast hourly shadows, and perhaps a decorative ring or two to hold the parts together, without regard to their correct positions. On the other hand, sometimes the mythological aspect is all that remains. For example, this figure of Atlas holding up an armillary sphere in Rockefeller Center in New York City functions as a narrative sculpture only. It lacks a central shadow-casting gnomon and was never intended as a dial.

The armillary is a type of equatorial sundial, but it was also one of the earliest astronomical instruments as well as a sacred divination tool of the Chinese, and it represented a summary of historical and philosophical views of the cosmos. We hope that you will remember its past days of glory while you watch its present hours of sunlight.

Sculpture of Atlas holding up an armillary sphere in Rockefeller Center, New York City. This celestial sphere does not function as a sundial, since it has no gnomon. It was created by sculptor Lee Lawrie (1877–1963) and installed in Rockefeller Center in 1937.

VI
The Mysterious Dilemma of the Analemma

Why Is Sunlight Disappearing from Your World?

World globes used to have a large figure eight in a conveniently empty part of the Pacific Ocean. Now many of them no longer carry this sign, and its name is equally obscure. You may not even be able to find the term *analemma* (from the Greek word for sundial) in many popular dictionaries nor in several of the major encyclopedias.

The analemma is a chart telling both the sun's daily declination (the altitude or position north or south of the equator) and also the equation of time (the daily difference between sun time and clock time.) The northernmost and southernmost tips of the analemma on world globes fall on the Tropics of Cancer and Capricorn, which were so named when the sun entered these constellations at the solstices, reaching its greatest distance from the equator—23° above it on June 21 and 23° below it on December 21. On these dates it stops and starts back toward the equator and the opposite hemisphere. Knowledge of the sun's declination is used in celestial navigation.

Sundial owners are most interested in its second source of information—the equation of time. Traditional dials carrying this graph on their dial faces tell viewers at a glance whether their dial is faster or slower than the clock on any day of the year.

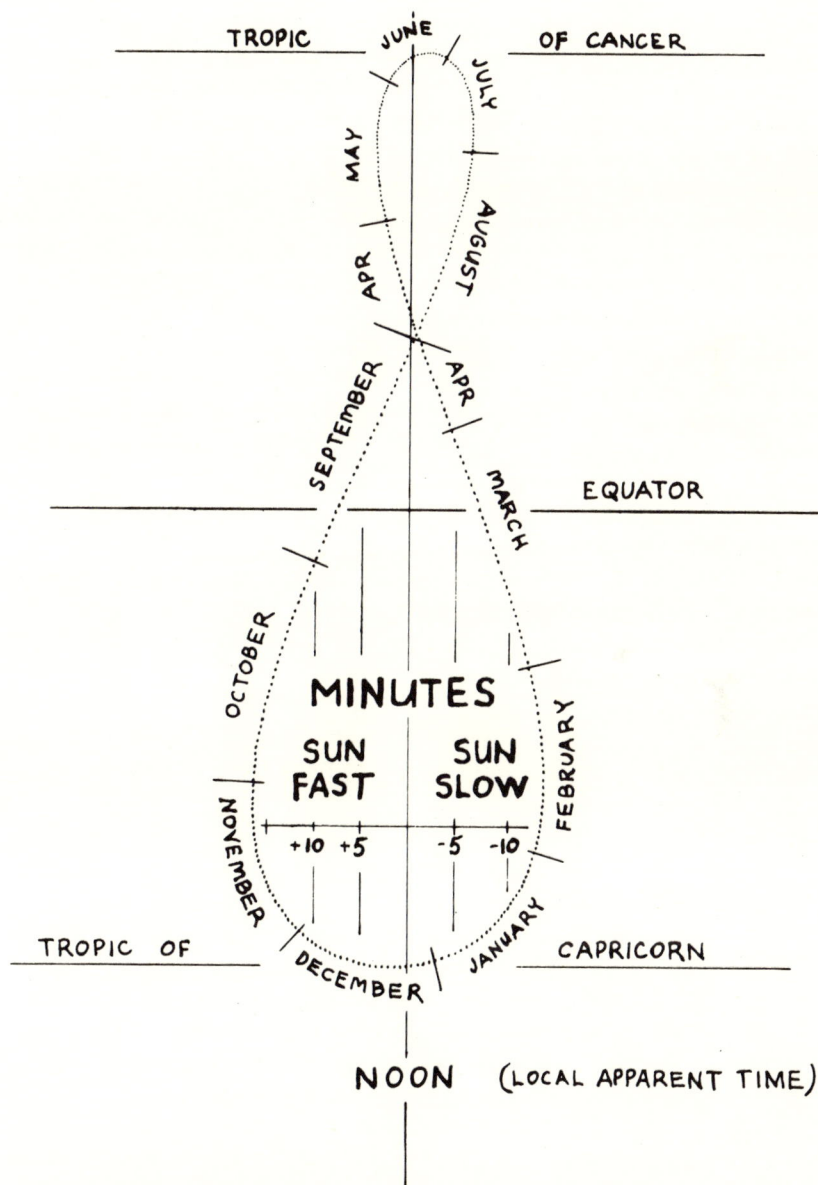

Analemma showing the equation of time.

This is how the analemma on a dial works: The center line running the long length of the figure represents the twelve-o'clock solar noon position (local apparent noon). The elongated figure eight line with the dates on it marks noon according to standard clock time. You can see that sometimes this line comes close to the center and at other times pulls away. This lateral, or sideways, distance represents the minutes of variation between the two systems of solar time and clock time. The dates when the difference is greatest are November 1, when the dial is a bit more than sixteen minutes faster than the clock, and middle February, when the dial is about fourteen minutes slower. The larger the size of this graph, the easier it is to mark the variations down to the seconds. The small scale of your miniature equatorial dial here will only give you a rough approximation of the equation of time.

The Equatorial Dial with Analemma

Cut out and assemble plate 9.

EQUATORIAL DIAL WITH ANALEMMA

GENERAL INSTRUCTIONS

1. Remove plate 9 from the back of the book before cutting out parts.
2. Cut only on gray lines.
3. Score all dashed lines (___) on the printed, or top, side of the page. Score all dotted lines (...) on the blank side, or underside, of the page. Be careful not to cut through the paper. Score lines with either a pattern-tracing wheel (also called a ponce wheel) or with a dull, slightly rounded tool, such as a butter knife (without teeth).
4. Fold all scored lines back away from the side of the paper containing the score.
5. Use a straightedge as a guide for cutting (with an X-Acto blade or mat knife), scoring, and folding all straight lines for the most precise results.

ASSEMBLY

1. In the center of the gnomon, cut out a very small but clean hole (a small diamond shape works well) with a very sharp blade.
2. On each side of the gnomon, completely cut out the panels marked "REMOVE," as shown in illustration 1.

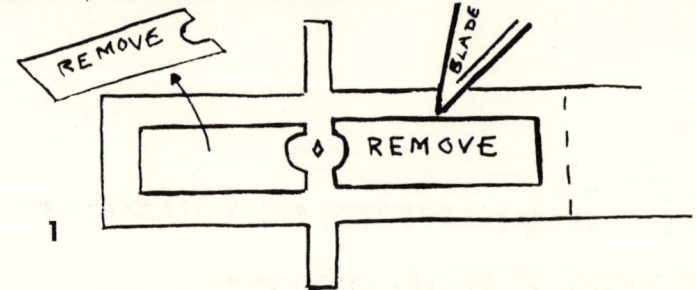

3. On the printed, or top, side of the page, score three lines—the two dashed lines next to the "N" and "S" marks and the line at your city's latitude number.
4. Leaving an additional 10° beyond your city's latitude number fold line, cut out the entire dial from the page. For example, if your latitude is 40°, score and fold on the 40° line, *but cut off* the strip at the 50° line, so as to give yourself a tab to glue onto.
5. On the blank, or reverse, side of the page, score and fold back the two dotted lines at the "6 AM" and "6 PM" marks and then fold back scored lines on the printed side.
6. Glue the dial band together as shown in illustration 2.

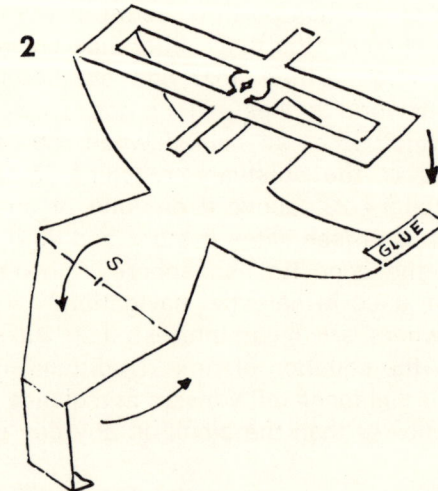

Model 9. Equatorial sundial with analemma. *Photo by Rosmarie Hausherr.*

7. Fold base along "S" and "N" scoring lines. Bring latitude strip up and glue the additional 10° tab to the back of the dial so that it aligns with the top edge of the curved hour lines, as shown in illustration 3.

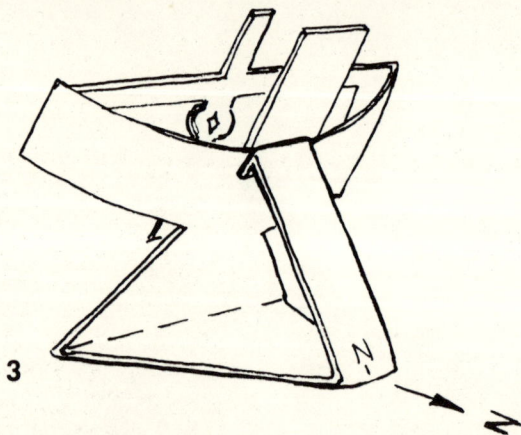

3

HOW TO USE YOUR EQUATORIAL DIAL WITH ANALEMMA

1. Position the dial so that the "N" and "S" marks align with the north–south meridian line. (Chapter 3 will remind you how to find true south.)
2. The shadow of the gnomon marks the hour.
3. *The analemma.* The figure eight in the center of the hour strip, called the analemma, is an added accessory here. If you do not care to make the longitudinal correction to your dial (described in chapter 4) to make it work, you can still use this cutout as an ordinary sundial without it. If you would like to use the analemma to make your dial into a solar calendar, or to make your dial time coincide with clock time, make your longitudinal corrections, and continue with the following instructions:
   a. The analemma pattern will be marked by the sunlight spot shining through the hole in the gnomon *at noon* each day (one point per day) throughout the year. When you mark the shape with monthly intervals (as shown on p. 59), the position of that light spot on the figure eight line shows the approximate date—thus giving you a solar calendar.
   b. The analemma is also a graph telling you "the equation of time," that is, the seasonal differences between clock time and solar time. Its shape is due both to the tilt of the earth on its axis and seasonal variations in the earth's rate of speed in its yearly orbit. When the sundial is slower than the clock, the sunlight spot falls on the left (or March) side of the center line; and when the sun and dial are faster than the clock, it falls on the right side. The lateral or sideways distance of the dot from the center line represents the minutes of difference between the two systems—sun and clock.

The noon line (local solar, or apparent, noon) is at the center of the hour strip within the analemma shape. The surrounding figure eight curve is roughly marked with the equation of time, not in exact minutes but by the labels "slow" and "fast" on either side of the center. This means that the dial is slower than the clock when the spot of projected sunlight from the pierced hole in the gnomon falls on the left (or March side) at *noon*, and that the sun is faster than the clock when the light is on the right (or September side) of the center line. This will only work if you make the longitudinal adjustment we talked about on page 44. Otherwise the figure eight projected by the sunlight spot throughout the year will fall to the left of center if you are east of the standard time meridian or to the right side if you are west of it (when you are facing south, or toward the sun at noon) and will not tell you correctly when the sun is fast or slow. Because of the small size of this model, we have not attempted to calibrate the minutes on it. The tiny lines and numbers would be so close together that they would not be accurate on this scale. What you have here is a demonstrational model rather than a precise astronomical tool.

You can mark in small lines for the monthly intervals to make a solar calendar here or else cut out the next model and put in the lines as you graph the analemma on a larger scale.

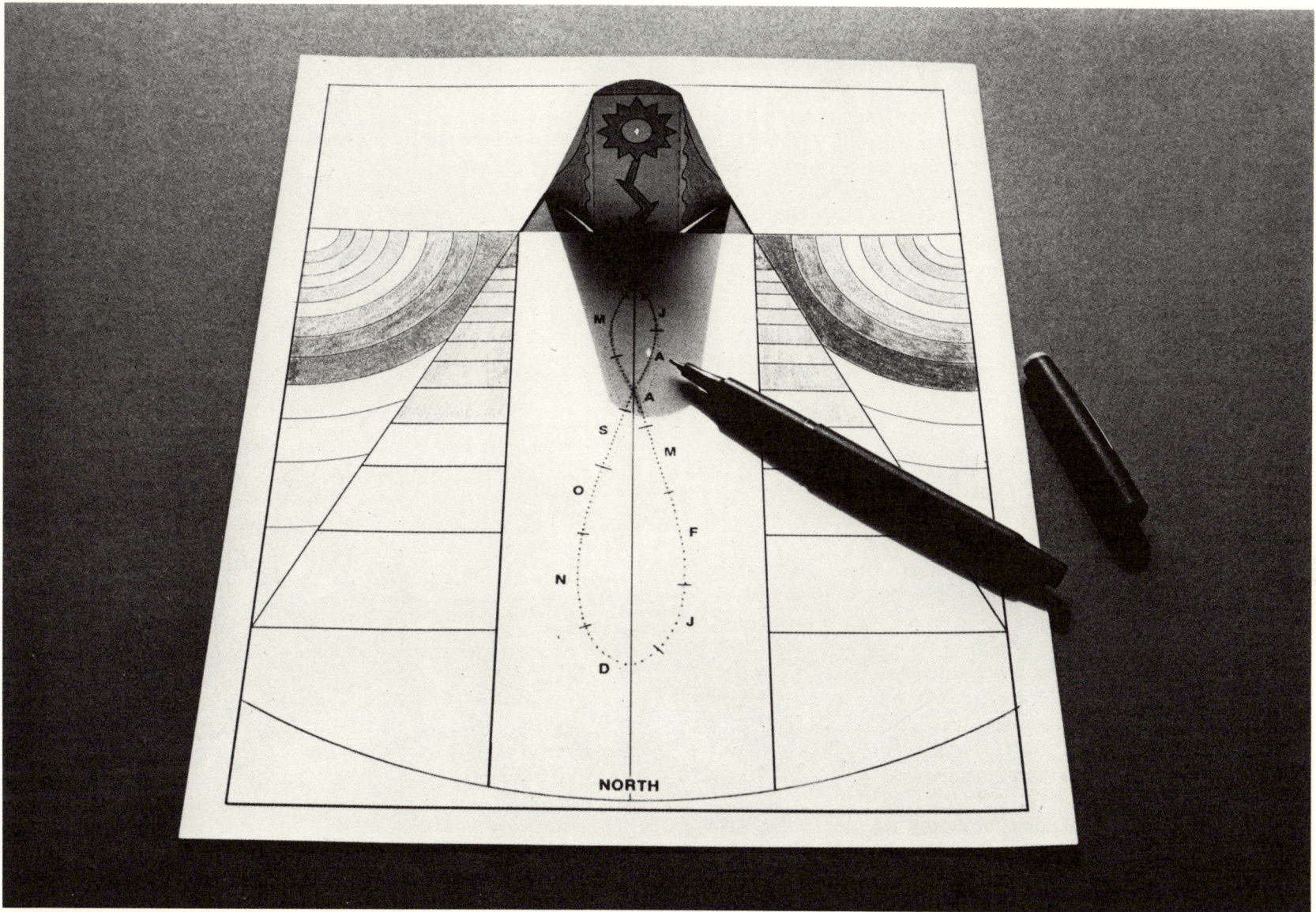

Model 10. Noon mark solar calendar. *Photo by Rosmarie Hausherr.*

NOON MARK SOLAR CALENDAR

The shape of the analemma was directly derived from drawings of projected sunlight. You can prove that for yourself with this model. The central section is empty so that you can fill in the dots and mark the exact dates where the noon sunlight spot falls on the horizontal surface each day. This figure eight when completed will give you a solar calendar for your own latitude. You cannot send this to someone else in another latitude, or it will not work accurately. That is why we couldn't graph a single typical analemma; each one varies according to your location.

Cut out and assemble plate 10.

GENERAL INSTRUCTIONS
1. Remove plate 10 from the back of the book before cutting out parts.
   Note! This entire sheet is used. Do not cut out and detach any individual sections from the plate except for the tiny center hole.
2. Cut only on gray lines.
3. Score all dashed lines (_ _ _) on the printed, or top, side of the page. Score all dotted lines (. . .) on the blank side, or underside, of the page. (Poke a pinhole through the last score dots at each end of the dotted line on the printed side to provide guide points when you turn the page over to score the blank side.) Be careful not to cut through the paper. Score lines with either a pattern-tracing wheel (also called a ponce wheel) or with a dull, slightly rounded tool, such as a butter knife (without teeth).
4. Fold all scored lines back away from the side of the paper containing the score.
5. Use a straightedge as a guide for cutting (with an X-Acto blade or mat knife), scoring, and folding all straight lines for the most precise results.

ASSEMBLY
1. In the center of the sunflower, cut out a small but clean hole with a sharp razor or X-Acto blade. (A small diamond-shape hole works well.) See illustration 1.

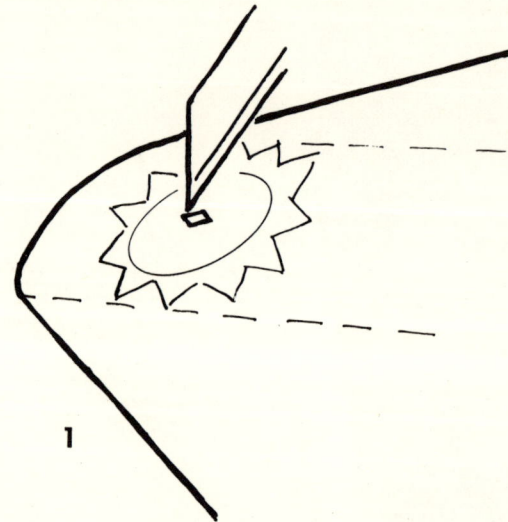

1

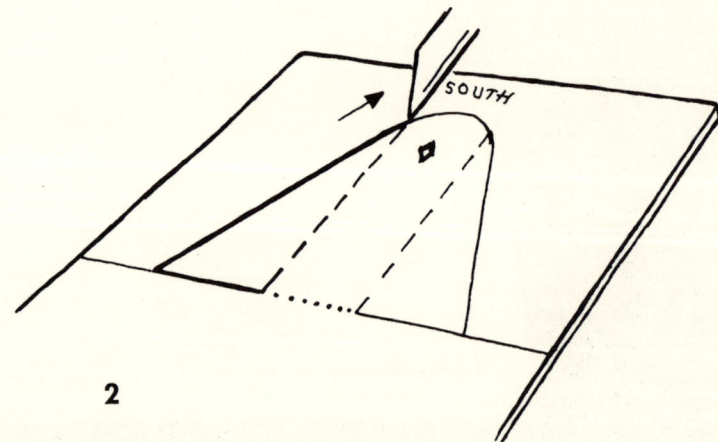

2

2. Cut on the gray lines, as shown in illustration 2.
3. On each side of the sunflower, score the dashed lines according to general instructions.
4. At the base of the sunflower, score the dotted lines according to general instructions.
5. Fold the sunflower upright and fold back its side supports as shown in illustration 3.

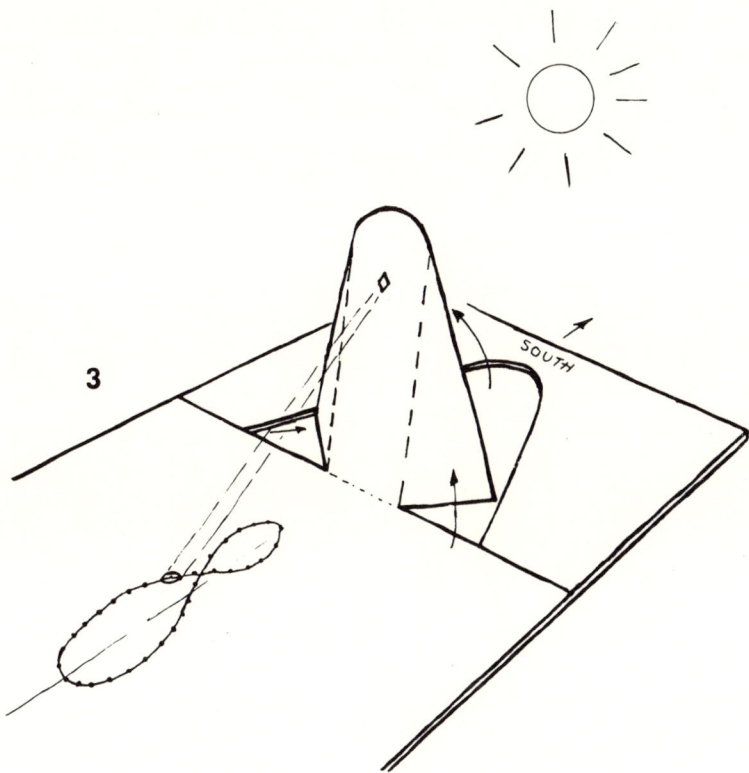

HOW TO USE YOUR NOON MARK SOLAR CALENDAR

1. Align the center "NORTH" and "SOUTH" marks on your model with the north–south meridian line.

2. Fasten the model securely in this position on your windowsill. Remember it cannot move for a whole year if your analemma drawing is to be as accurate as possible.

3. Each day, at the moment of exact noon Standard Time, or 1:00 P.M. Daylight Savings Time, mark the point and date where the sunlight spot falls on the dial table. If you do this every day of the year (on every sunny day, naturally) you will re-create that mysterious shape —the analemma—and also have an accurate noon mark solar calendar.

The Ceiling Sundial (Another Noon Mark Solar Calendar)

By the seventeenth century, every well-educated aristocrat was expected to know how to compute his own sundials. Whether they all actually made their dials by marking shadows on objects or walls from practical observation, or whether they all really mastered the branch of mathematics called *spherics* (three-dimensional, or spherical, trigonometry) and *gnomonics* (sundial making) is debatable. Many people must have acquired this information because it was necessary for navigation. Math and dialing were considered diversions of the nobility as well as being indispensable accompaniments to the then universal belief in magic, astrology, and other forms of occultism involving numerology.

Both Sir Isaac Newton and the architect Christopher Wren were reported to have made their own ceiling dials while students. We assume they computed their analemma dials mathematically— making trigonometric computations 365 separate times, one for each point on the analemma—for precisely their own latitudes, and this was certainly time consuming for a student without the convenience of our electronic pocket calculators.

An analemma solar calendar–ceiling dial, also called a "reflective" dial, is actually very easy to make without mathematical knowledge of any kind, although the process will take you an entire year. If you glue a half-inch or smaller piece of mirror to the inner part of your windowsill (preferably in a room that gets daily sunshine all year round), it will project a spot of sunlight on your ceiling. This light spot will trace out a complete analemma

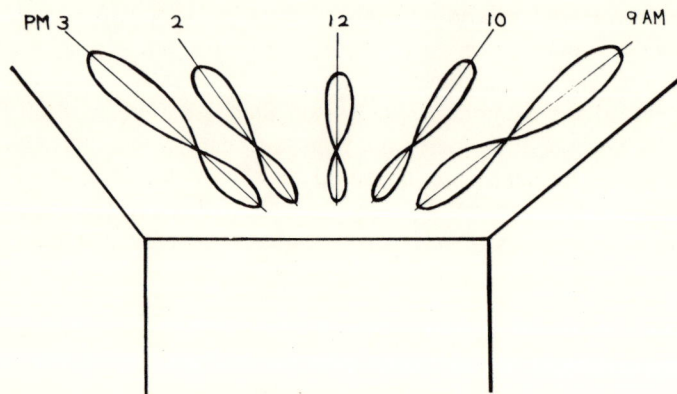

Ceiling analemmas for different hours of the day.

during the year if every day at exactly the same Standard Time you place a dot on the ceiling in the center of the sunlight spot. If you want to make it a calendar, include the date also. You can get a separate analemma figure for each of many hours of the day, if you mark the projected light spot every day at that hour.

We only marked the noon hour on our ceiling, since we decided that multiple hourly markings would be inconvenient. Bob's father gave us an adjustable date stamp to mark the calendar. Remember *you must place your mark at the same time every day.* Also make sure that you compensate for Daylight Savings Time by marking the dot an hour later during that period or else you will have a sideways displacement on your graph. If you do not date the marks, you must provide some code (for example, with different colors, shapes, or arrows) for yourself to differentiate the parts of the path that are traveling north on your ceiling (from June to December) from those dots going south (from December to June). Otherwise when the lines start converging in April and September, you may end up connecting the wrong dots. You can also color code each month or season, or put up two dots instead of a single one on the first day each month to calibrate the calendar (instead of, or in addition to, writing the dates on it).

We started our analemma last fall on the ceiling of our small kitchen. By December it had already moved down the north wall. We found out later that this was because we had put the mirror

in the wrong place for the particular size of this ceiling. We could have avoided the problem by placing our mirror piece up on the top sash of the window, instead of gluing it to the lower window sill. The farther away (or lower) the mirror is from the ceiling, the more it magnifies the analemma and the fewer different hour graphs will fit on your ceiling. If you start it in a large room or if you raise your mirror (keeping it horizontal), you should have no problem keeping the figure eight on the ceiling.

We were amazed to find out how fast the apparent sun actually travels. On January days when we'd been about ten minutes late for our marking, the sun traveled nearly a foot on the ceiling from where we knew the mark must have been exactly on the hour. So if we're more than a minute or two late in getting our dot up, we leave that date blank, and we do the same if it's cloudy and there is no sunlight spot to graph. Other gaps arise from vacations or

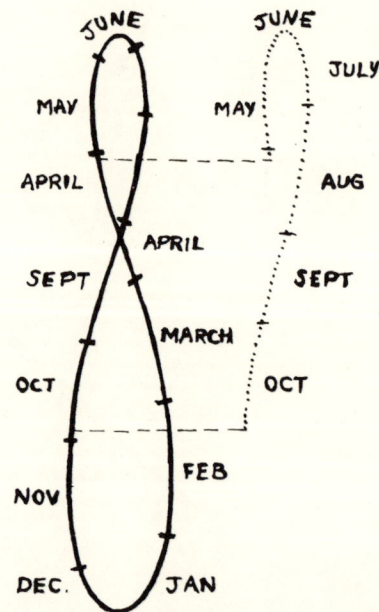

Analemma with sideways displacement for Daylight Savings Time.

66

days we are away from home for any reason at the marking hour. For short intervals we've missed, we merely make an estimated line through the preceding and following dots. On longer time spans it's problematic—we'll try to pick up those dots next year. Our photograph here shows not only sticker dots representing the center points of the light spots but also pencil marks outlining the lit shape. We used a tiny quarter-inch-diameter dot, although you can also get larger ones (three-quarter-inch) in many colors. Those are removable, self-adhesive labels, and they are sold in stationery stores carrying business supplies.

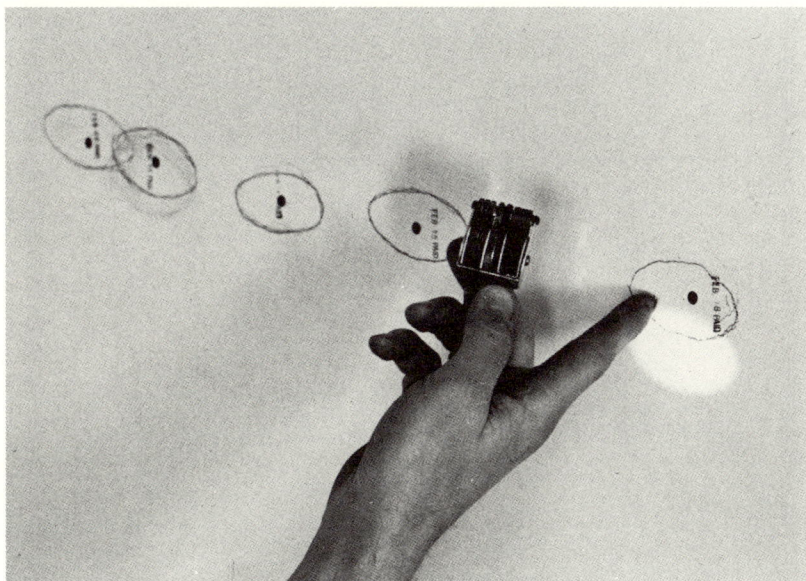

Section of our ceiling dial showing sticker dots, date stamp, and pencil outlines of the light spot. *Photo by Rosmarie Hausherr.*

The Floor Dial

You can put your analemma on the floor without the aid of a mirror by cutting out a small hole in the center of a piece of cardboard and taping it on the windowpane so that a beam of light projects through it onto the floor. Proceed with the marks the way you did with the ceiling dial. Old noon mark analemmas were sometimes set into the floor tiles as permanent timekeepers in centuries past.

If you do not have a southern window with the maximum number of hours of continuous sunlight per day, you can still make a limited single-hour dial. Mark the light spot on the hour when the sun *does* come into your room. If your room has several windows with different exposures, you might mark two analemmas at the same time using reflected light spots from each window. You can differentiate between them by different-colored or -shaped sticker dots. It will be interesting to see what happens if the lines cross. If you do all three dials—two ceiling models and one floor model—you'll have some marvelous graphic designs in your room.

You will also find that during the year, the shape traced out on your ceiling varies from that of the analemma you mark on your small paper model. This is because your ceiling graph is distorted by its oblique angle of projection onto the ceiling.

If you are only ambitious enough to mark the sunlight spot at the same minute of the same hour two times out of the year at days of nearly opposite solar positions (for example, late in June and late in December), you can still get a single-hour sundial by joining the end marks of the amalemma figure with a straight line. However, this line shows *only* the solar hour (or what is called local apparent time), not clock (or mean) time.

The dials we have introduced in this chapter are also called *spot dials* because they use a spot of light instead of a shadow to mark the dial surface. Your three equatorial dial variations (plates 4, 5, and 6) also work on this principle.

There is more than one meaning for the term *analemma*. Some writers refer to it as an astronomical-mathematical instrument with a movable *cursor* (Latin for "runner"), others think of it as a kind of simplified armillary sphere, and still others consider it a technique used to solve spherical trigonometric calculations. The analemma figure resembles the Greek symbol for infinity (∞), although we can find no references to any possible connection. The Roman architect-engineer Vitruvius mentioned the analemma in his writings in the first century B.C., and by the second century A.D. the astronomer Ptolemy was reported to have used an analemmatic device on his armillary sphere.

Dial enthusiasts of the seventeenth century developed another type of sundial from knowledge of the analemma. This dial differed from other dials because its gnomon was movable and did not point toward the north celestial pole, as did most other dials, nor did it necessarily chart the figure-eight shape. It was often mounted along with a stationary gnomon on a horizontal dial table and was used when the direction of the north-south meridian was unknown. By turning the table to make both dials tell the same time, you could position it correctly. Although we have not included a cutout model of this dial, you can see a simplified version in this drawing.

Why is the analemma disappearing from our reference books? The replacement of solar navigation tools by radar, computer, and other devices, as well as the disappearance of many sundials, may have something to do with its demise. It may be resurrected soon again when we start using solar power and want to know all about the sun's movements.

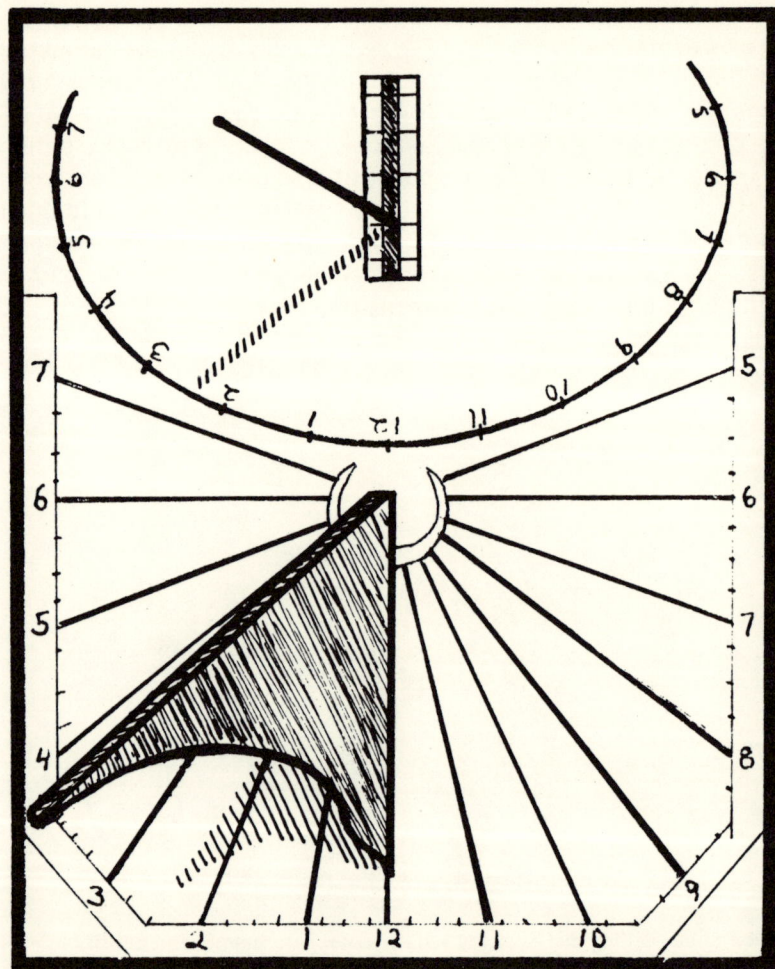

Analemmatic dial with movable gnomon mounted with stationary horizontal sundial.

VII
Marking the Time
(On the Meridian)—
East–West Vertical Meridian Sundials

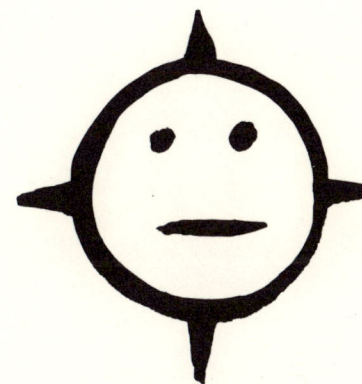

**Sun symbol
American Indian**

These next two models (plates 11 and 12) are design variations of the large-scale sculpture planned for a college mall. These dials belong in the gnomonic class called *east–west vertical meridian dials* (also known as *vertical direct east-and-west dials*) and are so named because their hour tables (or plates) are vertical to the ground, and face the east and/or the west and because their axes point toward the true south meridian line. This type of dial was often attached to European building walls facing either east or west with a flat sheet or bar as a gnomon attached to the 6 A.M.– 6 P.M. line. The custom probably began during the Renaissance, since that's when equal-hour dials became known throughout the West.

This dial was thought to have a Greek ancestor, although we have no direct reference to it by this name. A famous Greek building, the Tower of Andronicus (now called the Tower of the Winds), which was built in Athens in the first century B.C., had sundials added to all of its eight sides some years after it was completed. The dials on its eastern and western faces may have been of this variety.

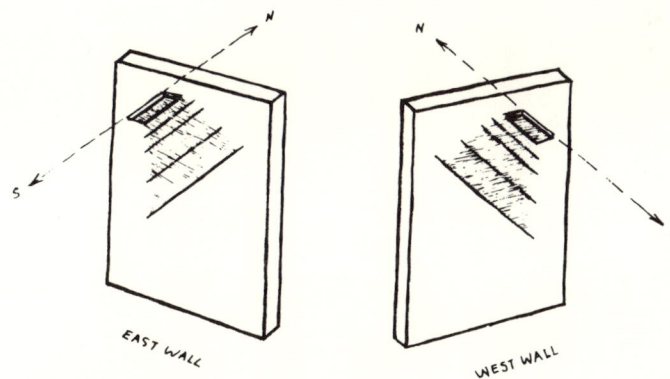

East and west walls showing vertical meridian dials.

69

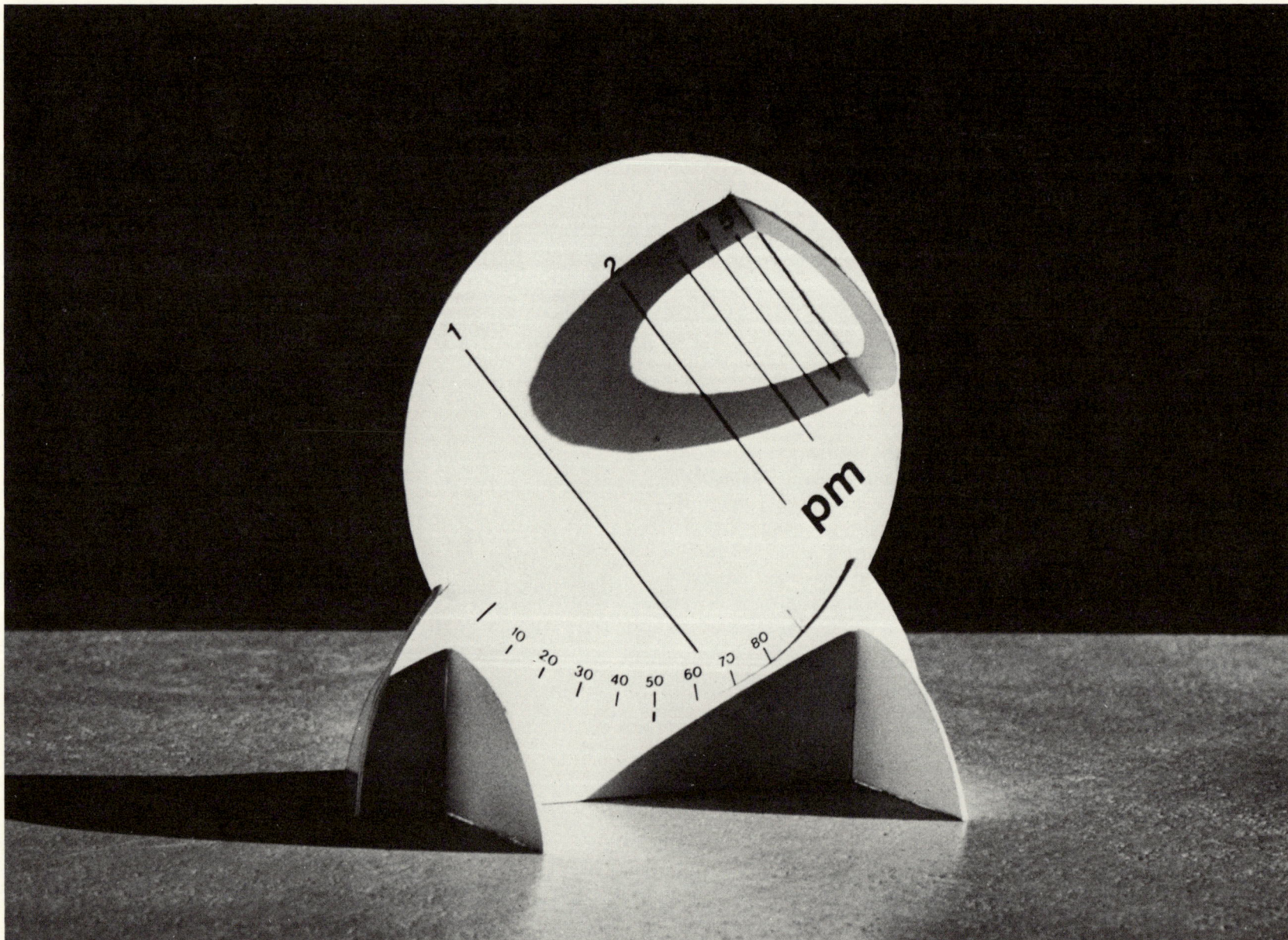

Model 11. "The Double Circles" (east–west vertical meridian dial). *Photo by Rosmarie Hausherr.*

"The Double Circle"

Cut out and assemble plate 11.

"The Double Circle" is a further development of *The Tower of the Sun*, described on page 4. Your own paper version here simplifies the form to its bare essence. The towering stem has been replaced by a horizontal base, which allows the dial to be adjusted to any latitude. The original *Sun Tower* was stationary and was made specifically for the latitude of its location. Its dial face (reminiscent of the silhouette of some spaceship service modules) is here transformed into a magic sun circle.

Bob chose to work with this type of dial in his dial sculpture project because the major avenues of the mall upon which the commissioned piece was to stand ran east and west. He wanted a dial face that would be especially visible from these two directions. Bob altered the ancient idea of using a building as the dial face by creating a free-standing dial—in other words, by eliminating the building between the faces.

"THE DOUBLE CIRCLE" (AN EAST–WEST VERTICAL MERIDIAN DIAL)

GENERAL INSTRUCTIONS

1. Remove plate 11 from the back of the book before cutting out parts.
2. Cut on gray lines (except where indicated differently).
3. Score all dashed lines (_ _ _) on the printed, or top, side of the page. Be careful not to cut the paper. Score lines with either a pattern-tracing wheel (also called a ponce wheel) or with a dull, slightly rounded tool, such as a butter knife (without teeth).
4. Fold all scored lines back away from the side of the paper containing the score.
5. Use a straightedge as a guide for cutting (with an X-Acto blade or mat knife), scoring, and folding all straight lines for the most precise results.

CUT OUT PARTS

1. Cut out A and A', including the slots on each at 6 A.M. and 6 P.M.

lines. Glue A to A' back to back so slots and latitude numbers align. See illustration 1.

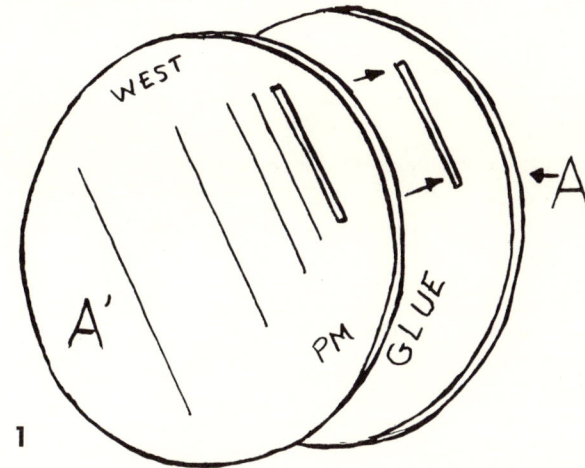

1

2. Cut out B *outside* gray line. Score and fold in half along line b–b'. With halves together, trim B *on* gray line, cutting out slots so they align. Glue halves together.
3. Cut out C *outside* gray line. Score and fold in half along line c–c'. With halves together, trim C *on* gray line, cutting out slots so they align. Cut C in half along dashed line c–c'. Do *not* glue halves together.
4. Insert and glue B between both cut halves of C, as shown in illustration 2.

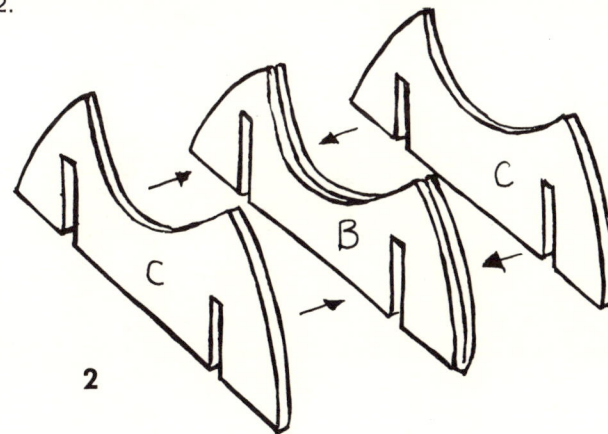

2

5. Cut out D and E *outside* gray lines. Score and fold each in half along their center lines. With halves together, trim D on gray line, cutting out slots so they align. Repeat process with E. Glue halves of each piece together. Do *not* glue D to E.
6. Cut out circle F. You may choose one of two methods here:
   a. Either cut out each circle, F and F', separately and glue them together;
   b. Or cut a large enough rectangle around one half of F, so that you can fold the pattern piece over to cut out one double-thickness circle instead. Glue both circles together.

ASSEMBLY

1. Insert D and E into bottom of now joined B–C base.
2. Insert dial face A–A' into center slot on top of base, as shown in illustration 3. (Do not glue.)
3. Insert F–F' halfway through slot in A–A' so that it is at right angles to the dial face A. See illustration 3. To check that it is at right angles to dial face, cut out and use right angle guide, piece G.

Note: If you wish to disassemble this dial for carrying or mailing, do not glue these parts permanently. Otherwise you can glue D and E to base B–C, and F–F' to dial face A, at the junctures between parts.

HOW TO USE THIS DIAL

1. Rotate A in slot of base until your latitude number lines up with the center mark on the base.
2. Position dial so that the edge of A (the end with F inserted) points south, as shown in illustration 4.
3. The sun shines through F, casting an elliptical light shape onto A. The forward edge of the *inner*, illuminated ellipse tells the time as it crosses the hour lines. As the sun rises in the morning, this

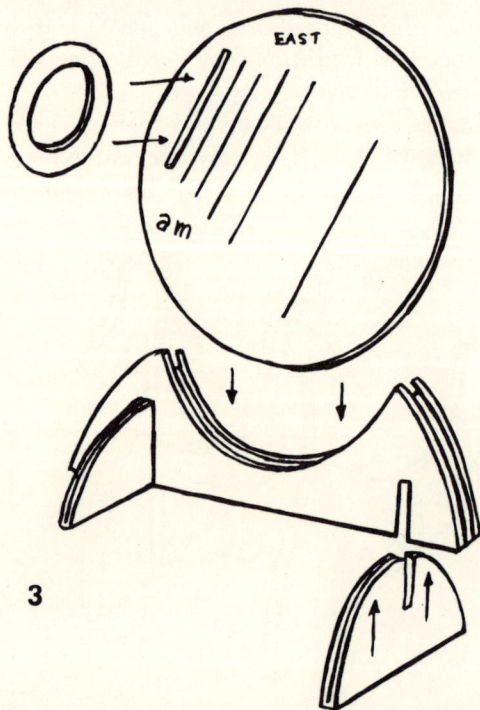

3

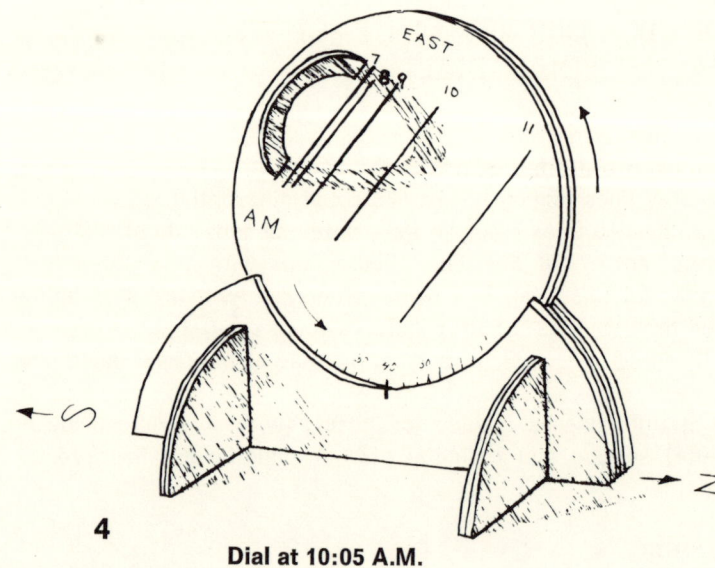

4

Dial at 10:05 A.M.

luminous shape elongates more and more on the eastern side until it disappears at high noon. Then the light shape gradually reforms and recedes back again up the face on the western side in the afternoon.

Note: For Daylight Savings Time, add one hour to time shown on dial.

The Pierced Gnomon

Most common modern sundials tell time by casting a shadow from a straight-edged gnomon onto the hour lines of a table or face, but a spot of light is equally if not even more dramatic as a marker. You have seen on the equatorial dial variations on plates 4, 5, and 6, and the noon mark solar calendar on plate 10 that small beams of light have relatively harder, more intense, edges, or margins, than do shadows.

The Chinese used *pierced gnomons* (called the "eye" of the dial) by the fifth century B.C. and the Arabs by the tenth century A.D. to make their dials more precise. The diameter of a small projected sunlight spot could be measured and bisected in order to find its center. That center point would tell the most accurate time. Post-Renaissance Europe also used light spots for noon marks that they projected onto their southern walls, just as you might have used a piece of cardboard with a hole in it on your window to cast an hour mark on your floor.

A hole cut into a shadow-casting gnomon to enable sunlight to pass through it is called a *nodus* (from the Latin word for "knot"). The circular cutouts on Bob's east–west vertical meridian dial gnomons are an expanded interpretation of this nodus idea. They form a type of extra-large dial eye that projects an elliptical light spot with a surrounding dark shadow from the remaining part of the gnomon onto the hour lines of the dial table.

"The House (for a Rising Sun)"

Cut out and assemble plate 12, "The House." Again, your paper dial is a scaled-down variation on an earlier large out-of-door

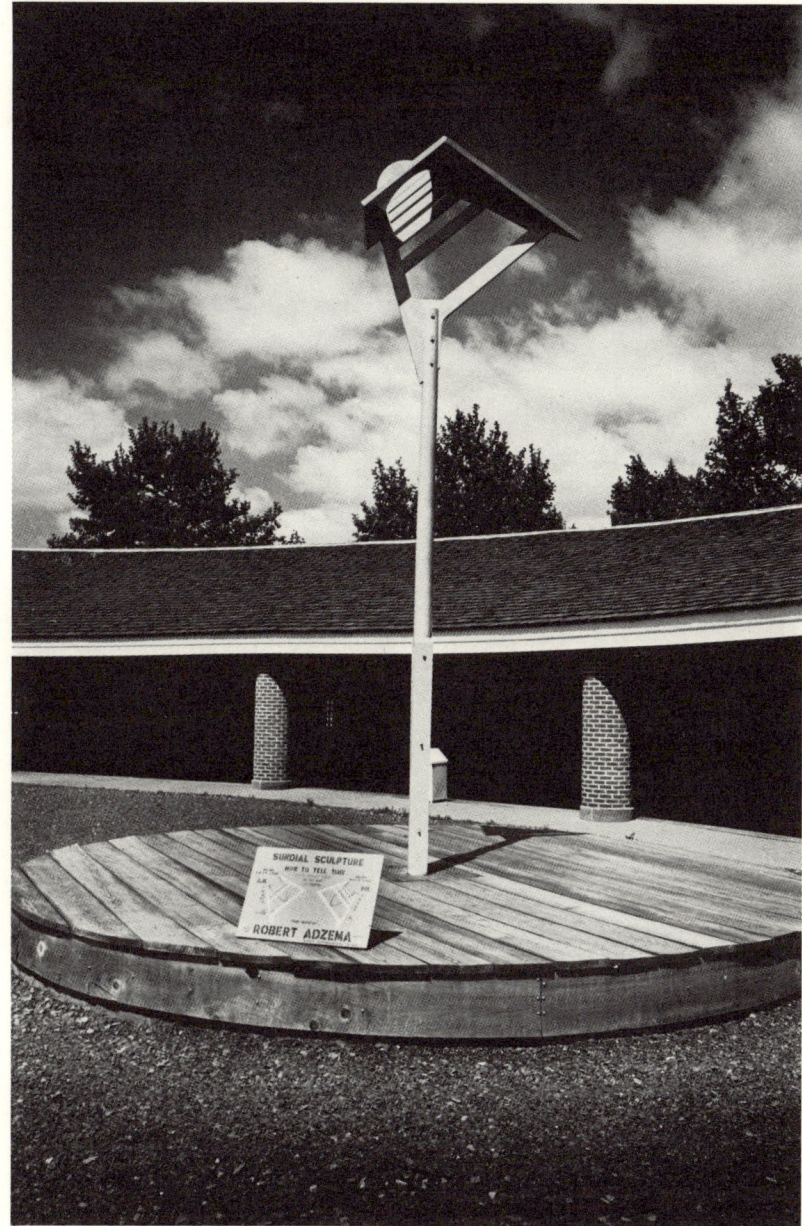

The House **(17 feet tall). Sundial.** *Sculpture by Robert Adzema*

Model 12. "The House" (east–west vertical meridian dial). *Photo by Rosmarie Hausherr.*

sculpture entitled *The House.* Here Bob has recalled the idea of a building dial with a less literal image of the building itself. He made the gnomon into a kind of roof that is pierced to project light.

"THE HOUSE" (AN EAST–WEST VERTICAL MERIDIAN DIAL)

GENERAL INSTRUCTIONS

1. Remove plate 12 from the back of the book before cutting out parts.
2. Cut only on gray lines.
3. Score all dashed lines (_ _ _) on the printed, or top, side of the page. Score all dotted lines (. . .) on the blank side, or underside, of the page. Be careful not to cut through the paper. Score lines with either a pattern-tracing wheel (also called a ponce wheel) or with a dull, slightly rounded tool, such as a butter knife (without teeth).
4. Fold all scored lines back away from the side of the paper containing the score.
5. Use a straightedge as a guide for cutting (with an X-Acto blade or mat knife), scoring, and folding all straight lines for the most precise results.

ASSEMBLY

Base (A)

1. Cut out base A, including the center slot.
2. Score and fold dashed lines according to general instructions.
3. Fold tab A over to side A and glue. Fold sides in and glue to side tabs. See illustration 1.

Dial Face (B)

1. Cut around both halves of B. Score on dashed line b–b′ and fold in half. Then (from one side) cut out interior sections, and then cut on outer gray outlines as well.

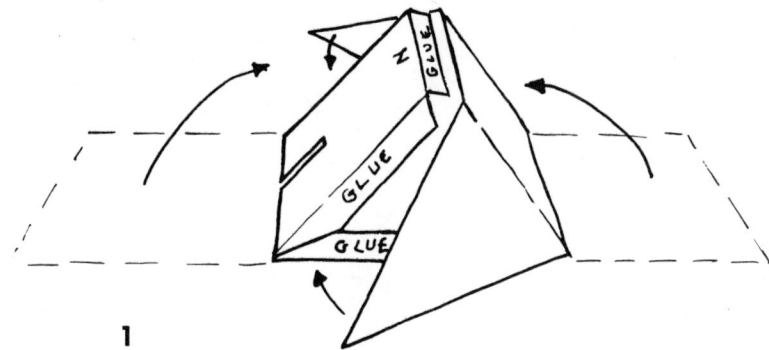

1

2. On the blank, or underside, of the paper, score tabs and fold according to general instructions.
3. Next, cut B in half along line b–b′.
4. Glue both halves together, *leaving scored tabs free of glue* and folded up.

Roof Gnomon (C)

1. Cut out C, including the center circle.
2. Cut slot in C wide enough to insert double thickness of B.
3. Score and fold dashed line according to general instructions.

Assembling All Parts

1. Place roof gnomon C over dial face B so that the half circle of B is centered and projects through hole of C. Glue short tabs on both sides of B to C.

2. Fold lower half of C (with slot) in to meet lower tabs of B and glue. See illustration 2.
3. Fold upper triangle of C down to meet upper tabs of B and glue.
4. Insert assembled dial unit B–C into slot of base A so that the bottom tabs of C are at the top and over the side of base A marked "NORTH."
   Glue bottom tabs of C to top of A only. See illustration 3.

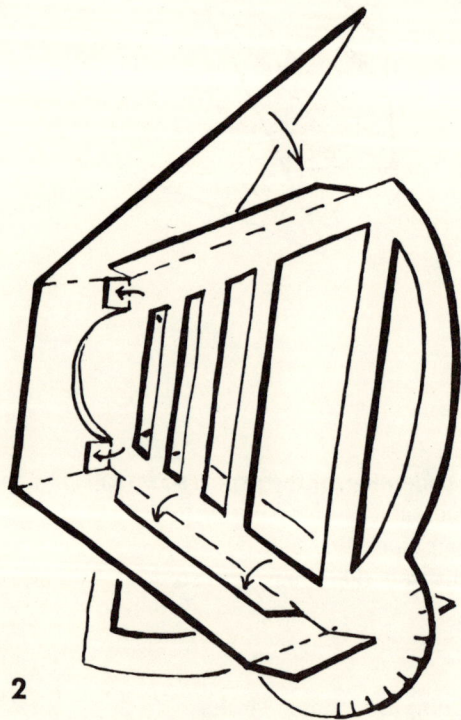

HOW TO USE YOUR EAST–WEST VERTICAL MERIDIAN SUNDIAL
1. Rotate dial unit with roof (B–C) in the base until the number of your latitude line aligns with the slot of the base. See illustration 4.

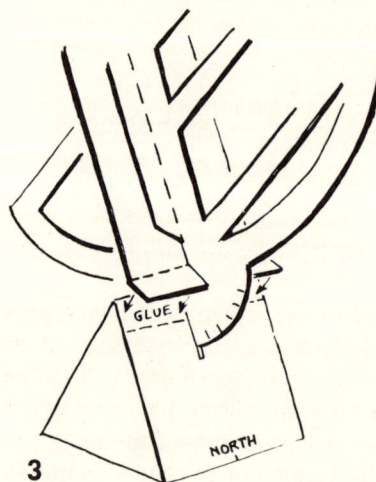

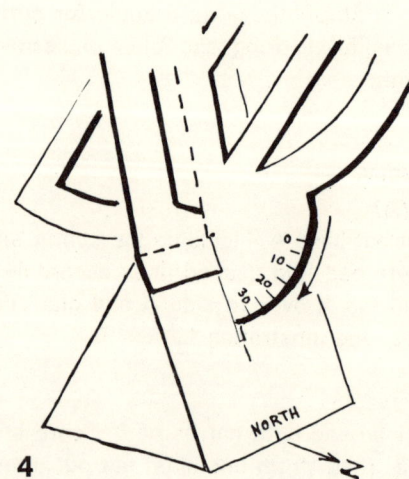

76

2. Position the dial so that the "SOUTH" and "N" marks align with true south and north directions.
3. Read the time at the forward edge of the *inner*, illuminated ellipse as it crosses the hour bars. As the sun rises in the morning, this luminous shape elongates on the eastern side until it disappears at high noon. Then the light shape gradually reforms and recedes back again in the afternoon on the western side of the dial.

Note: For Daylight Savings Time add one hour to time shown on dial.

According to many spiritual traditions, transcendental ideas and the elements of nature are expressed in color and form. For example, in early Buddhist belief, earth is represented by a cube, water by a white sphere, fire by a triangular shape with either a square or round base (that is, a pyramid or a cone), air by a hemisphere with its base tilted upward like a cup. When Bob sought ways to incorporate the basic sundial elements of gnomon and dial face into his work, he considered the ideas that each shape brings to mind. He knew he wanted an umbrella-type form to shelter and darken the area receiving the heavenly light, and the hole in the gnomon of the earlier sculptures called up the idea of a circular doorway for a sheltering house. But what kind of houses have circular doors? Birdhouses. What do birdhouses have to do with sundials? Birdhouses are often placed on top of poles to keep animals from getting at them—making that pole a kind of substitute tree. Trees are sun-related images in both ancient and oriental thought (for example, the Tree of Life). So the triangular roof (representing fire) seemed appropriate on the yellow (the symbolic color of light) house. So there it was: the "house" for the sun.

We discovered later that the idea of a house for the sun comes from mythology. The Greek sun god, Apollo, supposedly lives in a palace on pillars; and according to Chumash American Indian lore, the Sun retires each night with his two daughters into a quartz crystal house in the upper world, which is filled with every variety of tame animal.

Your own versions of these sculptures work in eastern and western windows as well as in southern ones. If you are using an eastern or western window, you will of course only be able to read the hours during that limited part of the day when the sun is shining through it. Lucky people with southern exposures get the greatest total number of sunlit hours a day; an eastern view will get light in the morning; a western one keeps the sun latest in the day. If you have windows on several sides of your home, office, or school room, you can use one of these different models for each window so that one dial picks up the light when a dial on another side goes into shade.

Now that you have the patterns and see how they can be altered to make original forms, why not try inventing some sun sculptures of your own? This collection of quick paper models set on top of split dowels shows six variations of the same dial. As long as you make the vertical plane pointing south exactly the same size as in your pattern, keep the gnomon the same size and perpendicular to it, you can mark your own dial table at intervals indicated in your model. You can change the outside silhouette shape or even make the surface form steps by creasing the paper, as shown here.

Six variations on the east–west vertical meridian sundial by Robert Adzema.

What kind of images does the sun make you think of? Another of Bob's models here resembles a rocket or arrow inside a circular gnomon. Arrows from the sun wheel have long been associated with rays of light. Connections between the sundials and other ideas pop up in the most unexpected places for us. For example, Bob found that another of his sun towers resembled two Chinese ideograms in the *I Ching*, the great book of oriental prophecy. He named the sculpture *Shêng* (see photograph on page 2) after one of these symbols. *Shêng*, or Ascending, has two parts—the earth and wood—which together represent trees and grass; nature rooted in the earth while growing higher. Another similar ideogram is *Ching*, or The Well, which combines the wood with water to signify a source of inner strength.

We hope that you will come to enjoy stretching the bounds of these patterns once you have mastered the elemental processes of putting them together. After all, the sundial shows you cyclic or unbounded time, and its mythical representative figures were symbols of eternal inexhaustible creativity. Let the sun inspire your imagination and allow you to create original dials yourself.

Chinese ideograms with sun symbols in them resemble silhouettes of Bob's large sundial sculptures on page 2.
a. *Shêng* or ascending;
b. *Ching* or The Well of Inner Strength.

VIII
Polar Sundials

The Horizontal Dial (A Polar Style Dial)

Cut out and assemble plate 13.

Here it is, the one which comes to mind whenever the word *sundial* is mentioned: the common horizontal dial sold in many garden-supply stores. You may even be asking yourselves, "Why didn't they put this model near the front of the book? It even looks simpler than many of the previous cutouts." Indeed, some dialing texts start with this, and then backtrack historically to show how it was developed. We have placed it here because this form was perfected in the West after centuries of dial making, and its construction involves not only understanding of celestial coordinates and terrestrial motions, but also knowledge of how to transfer lines of orthographic projection (that is, lines drawn on a surface at an angle to the plane they are projected from).

This dial is "horizontal" because of the position of the dial table, although its generic or family name—polar style dial—comes from the fact that the top shadow-casting edge of the gnomon, called the *style*, points to the north celestial pole. Polar pointing styles are not limited only to horizontal dial tables. The table can tilt in any direction as long as the style remains at the pole. But the relationship between the table and style is fixed. You cannot change the angle between them. To adjust a polar style dial for any latitude, you can tip the whole device up and down—but you cannot move any single part of it. If you tip the dial table alone while keeping the style pointing to the pole, you must redraw the hour lines.

HORIZONTAL SUNDIAL WITH UNIVERSAL WEDGE

GENERAL INSTRUCTIONS

1. Remove plate 13 from back of the book before cutting out parts.
2. Cut only on gray lines (except where indicated differently).
3. Score all dashed lines (_ _ _) on the printed, or top, side of the page. Score all dotted lines (. . .) on the blank side, or underside, of the page. (Poke a small pinhole through score dots at each end of dotted lines on printed side of the page so you will have guidelines to score on when you turn the sheet over.) Be careful not to cut through the paper. Score lines with either a pattern-tracing wheel (also called a ponce wheel) or with a dull, slightly rounded tool, such as a butter knife (without teeth).
4. Fold all scored lines back away from the side of the paper containing the score.

Model 13. Horizontal sundial (with universal wedge). *Photo by Rosmarie Hausherr.*

5. Use a straightedge as a guide for cutting (with an X-Acto blade or mat knife), scoring, and folding all straight lines for the most precise results.

ASSEMBLY OF DIAL (PARTS A AND B)

1. Cut out gnomon A.
2. Score all lines where indicated according to general instructions.
3. Fold up A into a triangular form. Glue tab A inside side A, as shown in illustration 1. Set A aside.

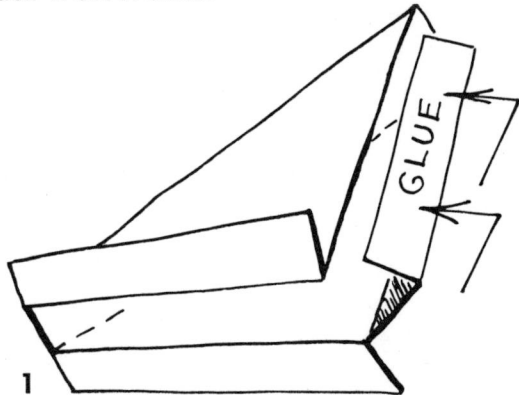

4. Cut out slot marked "REMOVE" from Dial Table B. Cut out Dial Table B according to your latitude number:
   a. If your latitude number is at or within 5° of 40°, cut out B on the gray lines and score the dashed lines. Fold the scored borders and glue corner tabs inside the rectangle as shown in illustration 2.

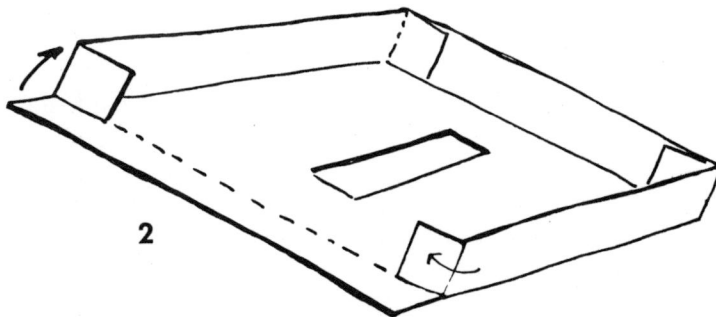

   b. If your latitude is *not* within 5° of latitude 40°, cut out B on the *dashed* lines, cutting off the outer margins and tabs (you will later glue this flat piece to the wedge).
5. Insert A from underneath through the cut-out slot of B so that the lower angle of the triangle faces south, as shown in illustration 3. Glue A to the bottom of B.

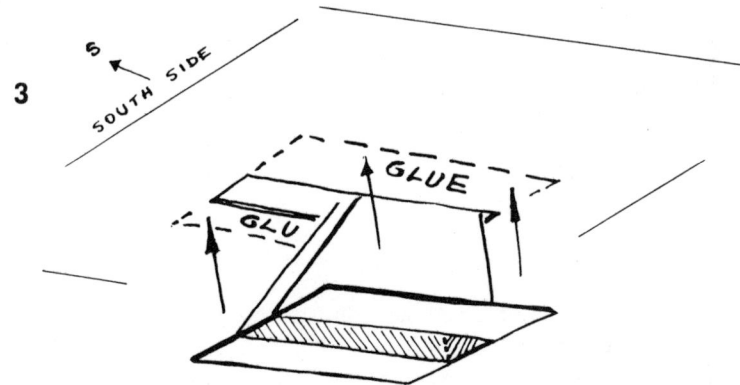

6. If your latitude number is within 5° of 40°, your dial is finished and you do not assemble the wedge. If your latitude number is *not* within 5° of 40°, you must assemble the wedge.

ASSEMBLY OF UNIVERSAL SUNDIAL WEDGE (C)

1. Cut out wedge C and trim to the numbered lines carefully according to the following instructions: Determine the number of degrees your latitude is more or less than 40°. Then cut on the lines corresponding to that number on all three lined sections of C. For example, if your latitude is 55° or 25°, each of which is 15° away from 40°, cut on line 15 on both the two triangular side panels and also the central rectangular piece, as shown in illustration 4.

83

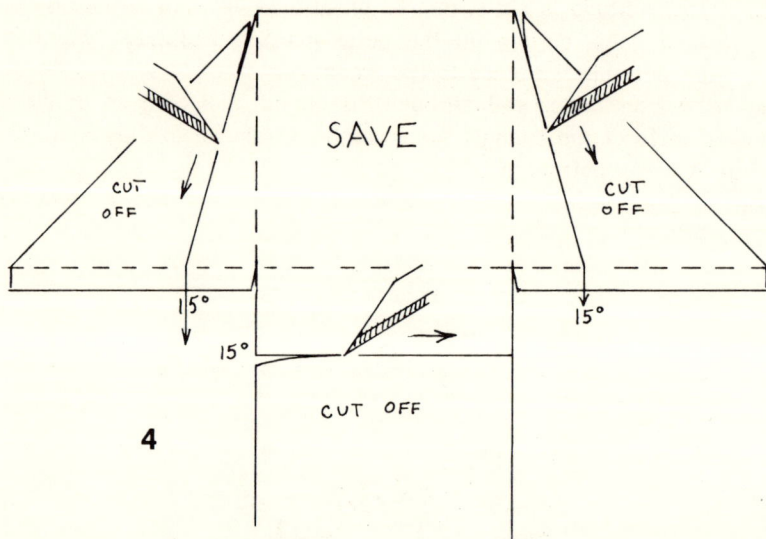

4

SAVE

CUT OFF

CUT OFF

15°

15°

15°

CUT OFF

2. Score all lines where indicated according to the general instructions.
3. Fold up the sides and glue the back of the wedge to the tabs of the side triangular panels, as shown in illustration 5.

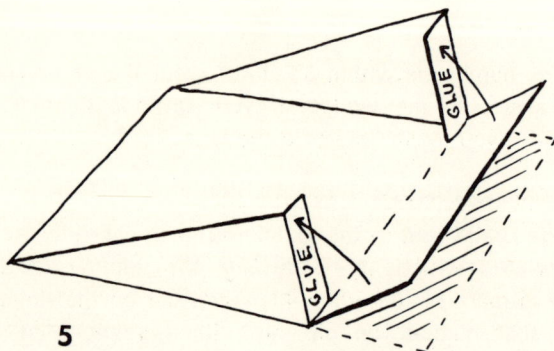

5

GLUE

GLUE

4. Glue assembled dial onto the wedge so that the height of the gnomon style equals the number of your latitude. For example, if your latitude is more than 40°, say 55°, put the high part of the wedge under the back of the table (at the twelve o'clock side); if your latitude is less than 40°, say 25°, put the high part of the wedge under the front of the dial (at the "SOUTH" side). See illustration 6.

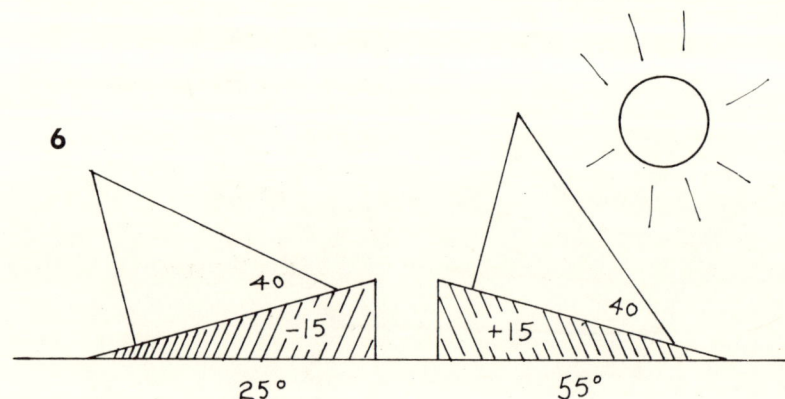

6

40

−15

25°

+15

40

55°

HOW TO USE YOUR DIAL

1. Position your dial so that the "SOUTH" mark on the table lines up with true south and the twelve o'clock mark lines up with true north.
2. Read the time where the leading edge of the shadow crosses the hour line.

The origins of this horizontal model are varied. A dial called the *polos*, used in ancient Babylonia, was reputed to have a polar style, but it was set inside a bowllike hemisphere. Eudoxus of Cnidus, a disciple of Plato in the fourth century B.C., was said to have made a horizontal dial with a vertical (instead of a pole-pointing) gnomon, which would have shown the unequal temporal hours. Finally, Aristarchus of Samos, a third-century-B.C. Greek astronomer, was credited with using a polar style gnomon on a horizontal dial table. Astronomers used the equal-hour divisions for their calculations although they stuck with the popular unequal hour system for civil affairs.

The first-known how-to instructions for creating this dial were written by the Arab mathematician-astronomer al-Battani (Albategnius) around A.D. 929, and the idea was brought to Europe by the Crusaders on their return from the Near East. These dials did not become popular until after the Renaissance acceptance of the equal-hour system, however, but once they were established, they replaced many of the older types showing unequal hours. That is probably one reason why we see so much of this type of sundial.

If you place your assembled equatorial wheel next to this horizontal dial, you may see a family resemblance between the two. Both of their gnomons and styles (or shadow-casting edges) point to the north celestial pole. What makes the construction of the horizontal model so different?

The difference is the way in which the hour lines are drawn onto the horizontal table. Dial makers had to develop a method for projecting equal hours of time that would be graphically unequal (that is, not equidistant, as on a circle). With your equatorial dial, the dial plane is always parallel to the equator so that the gnomon's shadow simply moves in a circle around it and the hours are all equal. But when you tip the dial face or table above or below that equatorial plane, the gnomon casts daily shadows of different lengths and shapes: They come out as ellipses, parabolas, hyperbolas, they even trace out a straight line on certain days of the year. To demonstrate to yourself how sunlit shapes project differently on variously tilted surfaces, merely hold up a pencil in the sunlight and use a piece of paper as a dial face—to catch its shadow. Tilt the paper with the shadow on it in many directions and watch what happens.

Longitudinal and Daylight Savings Time Corrections on the Horizontal Dial

The process of drawing projected shadow lines on a horizontal table is analogous to perspective drawing in painting (which also developed in the Renaissance). Perspective creates the illusion of space by projecting object shapes to one or more vanishing points (places in the background where straight lines converge). Think of the center point at the base of your gnomon as a vanishing point in a painting to which all the hour lines are drawn. Therefore, if you rotate the horizontal dial (horizontally), hoping to advance the hours for Daylight Savings Time, it will be like twisting the perspective system in a painting. So if you want to advance it for Daylight Savings Time, you must paste new numbers over the present ones or simply advance the hours mentally. The angle of the sun remains the same, only the hour number changes. However, *you must redraw all hour lines to make a longitudinal correction*. For example, if your dial is ten minutes fast (after you adjust for the equation of time), it is because you are east of your Standard Time zone line. Ten minutes is one-sixth of the hour space already drawn on your dial, so move the hour mark ahead one-sixth of the distance. If your dial is slow, move the mark in the reverse direction. After these adjustments, your hour lines will no longer be symmetrical but rather, will be shifted slightly to the right or to the left of the twelve o'clock mark. Remember that you still have the Standard Time numbers on the new lines; you need to add an hour to get Daylight Savings Time. Also remember that the equation-of-time adjustment is not affected by this change and your solar time will still be faster or slower than the clock according to particular seasons.

Secrets of the Sundial Makers

Popular belief has it that you have to know trigonometry to construct a sundial. Although this knowledge does help, you actually don't need it. All you need to know is a simple set of geometric construction methods and how to use a compass and ruler. These methods were kept secret by early dial makers because people skilled in the art of numbers were thought to have occult powers. They also found it financially lucrative to cloak their "trade secrets" in mystery. The Greeks considered the straightedge and compass to be instruments of the gods, so the sundial and its makers commanded great respect and at times were thought to be associated with the miraculous. The majority of people, on the other hand, knew only finger counting until after the Renaissance, and our Arabic numbers were banned in many countries before the fifteenth century. A fifteenth-century German merchant seeking the best advanced mathematical education for his son was told by a prominent university professor that German universities confined their curriculum to adding and subtracting. The Italian universities, on the other hand, also taught multiplying and dividing and were the only schools where the boy could get such an advanced education. Arithmetic that grade school children now perform in minutes meant hours or days of work for a medieval mathematician pushing some strung beads around his counting board or abacus. All of which helps to explain why this simple horizontal dial was so late in development even though the polar style dial was discovered many centuries earlier. After the flowering of mathematical achievements by the Greeks (ca. 600 to 300 B.C.), it took the Western world until the fourteenth and fifteenth centuries to awake from the intellectual shadow of the Dark Ages and complete the development of the sundial. The following three centuries became known as the Golden Age of Sundials, and the dial's decline came not with the invention and use of the clock, but with the Industrial Revolution in the nineteenth century when society's attitude toward time and work production had changed.

The Universal Sundial Wedge

The wedge for your horizontal dial (which you do not usually see on the mass-produced horizontal models) makes your dial adjustable for many latitudes. The commercial dials without the wedge don't really tell solar time, let alone approximate clock time (even with the seasonal and longitudinal corrections), unless their gnomons are cut at an angle appropriate to the latitude where they are to work. The horizontal dial gnomon (without the wedge) you just cut out and assembled was calculated for latitude 40° (the approximate latitude of New York City, where we live). If you don't live on or near this line, you will have to construct and use this wedge to tilt your style up or down in order to make it point correctly. The angle of the style must always be *equal* to

the latitude number of your city. For example, you need a 30° angle at latitude 30°, a 60° angle at latitude 60°, etc. This is the reason that all the sundials in this book have flexible construction with parts to cut off or to rotate to allow for your specific latitude. There is no part of your horizontal dial you can alter to make the gnomon work because you will disturb the relation of the triangle to the previously drawn hour lines on the dial plate. If you change that triangle in any way, you will have to redraw all the hour lines. This is why you must cut the wedge to accommodate your latitude and use it under the dial.

Sundials that can be tipped up or down to accommodate any latitude are said to be universal or portable dials. Remember that in cutting parts off your models, you make them permanent for your latitude only—as in the case of this wedge.

Your east–west vertical meridian dials, on the other hand, can be rotated constantly to any latitude as you travel. Remember, if you want to send someone who lives in a different latitude a horizontal dial (or any other dial that needs permanent alterations), don't cut it off for your own location but for their latitude instead.

The Polar Plane Dial

Cut out and assemble plate 14.

The polar plane dial is another variation on the polar-pointing style type. Here both the dial face as well as the style of the gnomon are parallel to the earth's axis. Compare this model with two or three other assembled dials, such as the wheel dial and horizontal dial, to see how the shadows are projected differently with the changed angle of the dial face. All of these three gnomon styles point to the north celestral pole. Note that on the polar plane dial there are no hour marks for six A.M. or six P.M. because the sun's rays are then parallel to the dial table plane and therefore cast no shadow on it. Before six A.M. and after six P.M. the sun is below that surface. On the horizontal dial the sun will shine on the dial table as long as it is above the horizon. With the equatorial wheel remember that the sun shines half the year on the bottom of the dial table. What we are observing is that although all these sundials have styles that are polar pointing, their names *horizontal*, *equatorial*, *polar*, *vertical meridian*, etc., all originate from the position of their dial tables. Now the dial family relationships will be obvious to you.

THE POLAR PLANE SUNDIAL

GENERAL INSTRUCTIONS

1. Remove plate 14 from the back of the book before cutting out parts.
2. Cut only on gray lines (except where otherwise indicated).
3. Score all dashed lines (___) on the printed, or top, side of the page. Score all dotted lines (...) on the blank side, or underside, of the page (on the printed side poke a small pinhole through the last score dots at each end of the lines so you will have guide points for scoring when you turn the paper over to the blank side). Be careful not to cut through the paper. Score lines with either a pattern-tracing wheel (also called a ponce wheel) or with a dull, slightly rounded tool, such as a butter knife (without teeth).
4. Fold all scored lines back away from the side of the paper containing the score.
5. Use a straightedge as a guide for cutting (with an X-Acto blade or mat knife), scoring, and folding all straight lines for the most precise results.

ASSEMBLY OF GNOMON (A)

1. Cut out A.
2. Score all dashed and dotted lines according to general instructions.
3. Fold up sides and ends; fold down tabs C and C' (which will later attach to B).
4. Glue tabs a and a' to sides a and a', as shown in illustration 1. Set gnomon aside.

ASSEMBLY OF DIAL TABLE (B)

1. Score all dashed lines on B according to general instructions.
2. Cut off table legs at the number equal to your latitude, as shown in illustration 2. If your latitude is 0° (at the equator), cut on lines b

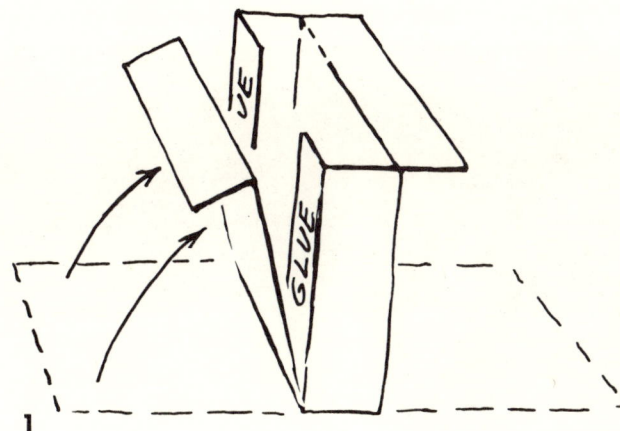

and b' to remove the rest of the table and legs. You do not need to tilt your table up. Insert gnomon to finish your model.

3. Cut out B, including the slot marked "REMOVE."

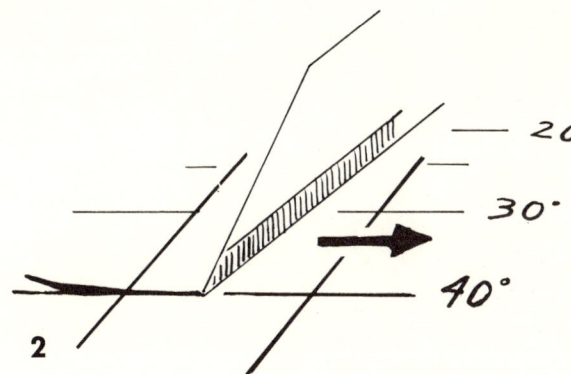

4. Insert gnomon A into slot on underside of B, gluing tabs C onto the blank side, or underside, of B as shown in illustration 3.

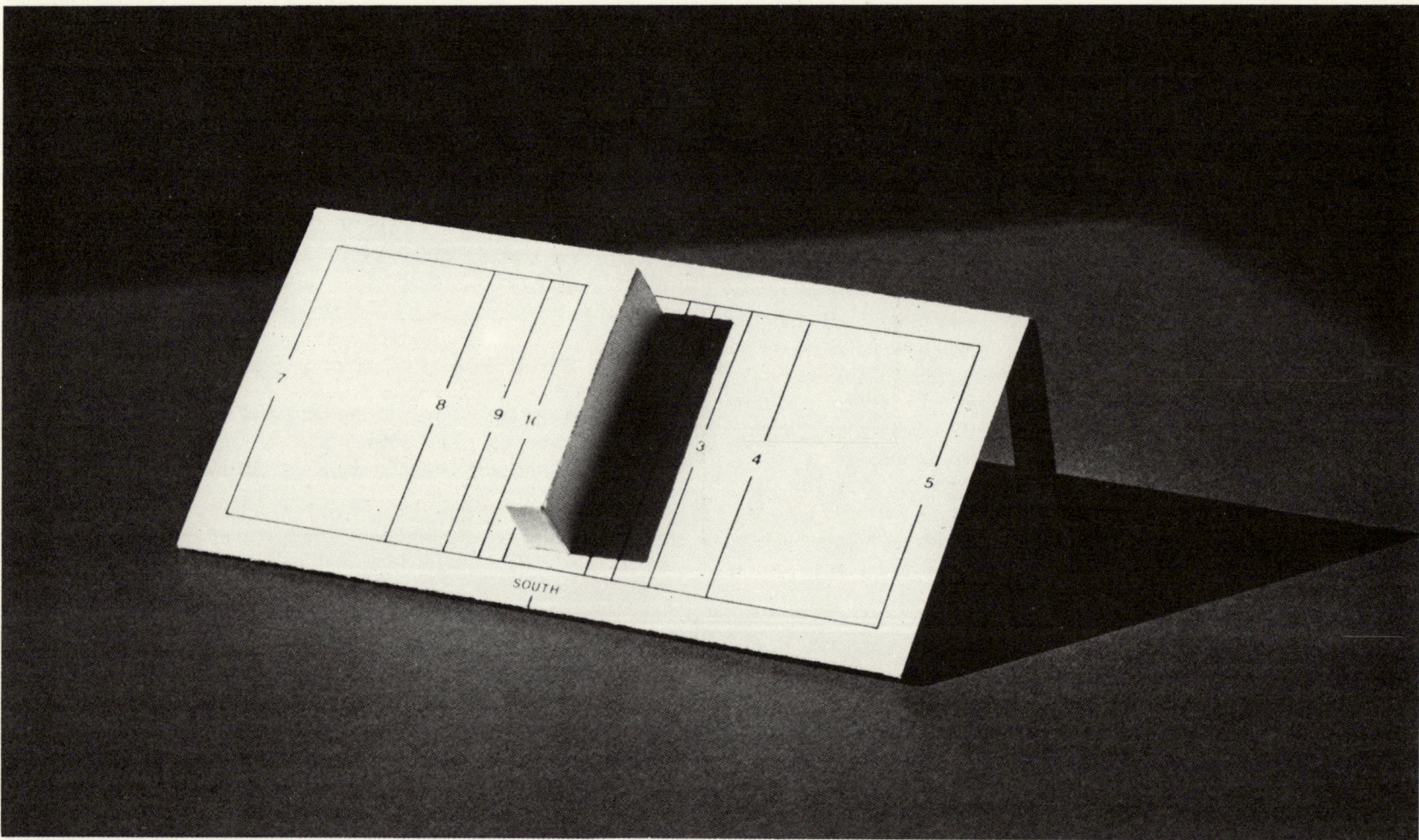

Model 14. Polar plane sundial. *Photo by Rosmarie Hausherr.*

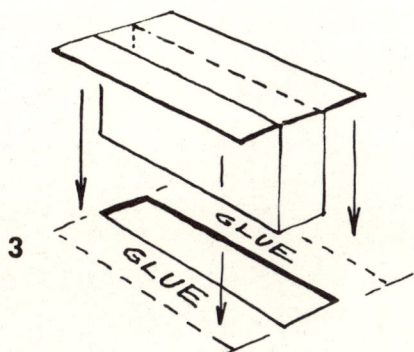

3

5. Fold B on all score lines: Fold down table tab on b–b' line. Fold up back legs to meet dial table tab as shown in illustration 4. Glue legs to underside of table tab.

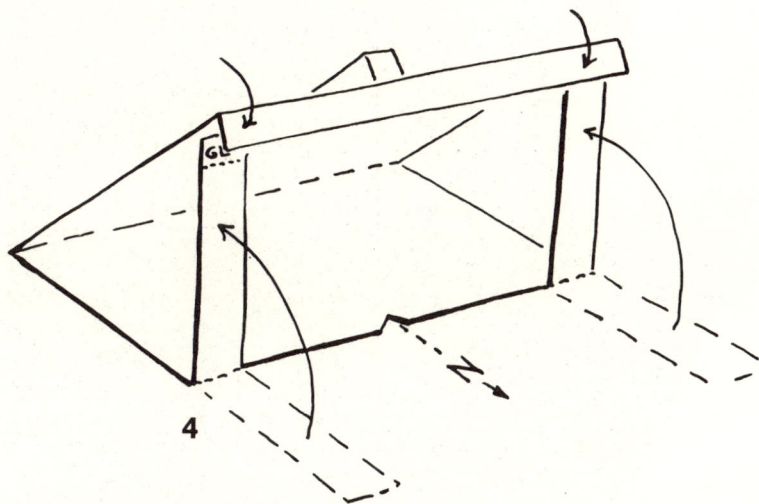

4

HOW TO USE YOUR POLAR PLANE SUNDIAL

1. Position dial so that the "SOUTH" mark near the gnomon and the triangular cutout notch on the dial bottom align with the north–south meridian line and the hour numbers face south.
2. Read the time where the edge of the shadow falls on the hour lines.

CUBE SUNDIAL

This probably looks pretty advanced unless you've already made both the horizontal dial and either of the east–west vertical meridian dials. You see, we've merely put both versions together and mounted them on different faces of the cube so that all the gnomon styles point to the north celestial pole. You have versions of the vertical east and west meridian dials on the eastern and western sides (without the pierced eye through the gnomons as on "The Double Circle" and "The House"), vertical direct north and south dials on the northern and southern faces, and a horizontal dial on top.

In addition, when you tilt the cube back to compensate for your latitude, the vertical dial facing south and the horizontal dial on top are said to "recline." That is, when a dial face or table leans back away from you as you stand in front of it, it reclines away from the zenith (the highest point overhead in the sky). It is then called a *reclining dial*, although this is not really a separate type of dial and the name is only due to the position of the face. If you tip a dial face toward you, it is correspondingly called an *inclining dial*.

As in the east–west vertical meridian models of the last chapter, this cube dial also probably had its origin as a collection of dials on different sides of buildings. When it was made free-standing, like your version, a joint was included in its pedestal to tilt the form for any latitude in order to make it "universal." During the Renaissance, dial makers displayed their virtuosity by making polyhedron dials with as many as ten faces. Of course, the sun will only shine on a few faces of any of these (including your own) and only for a part of each day. The southern vertical and horizontal dial faces will show time for the greatest number of hours.

Although there were block dials of various shapes in the ancient Near East (for example, in fourth-century-B.C. Egypt), these probably did not have gnomon styles pointing toward the north celestial pole. They merely cast the shadow of one projecting part onto another surface that was marked with hour lines. Our cube is a polar style dial; all the gnomon styles are parallel to the earth's axis and so are described as polar pointing.

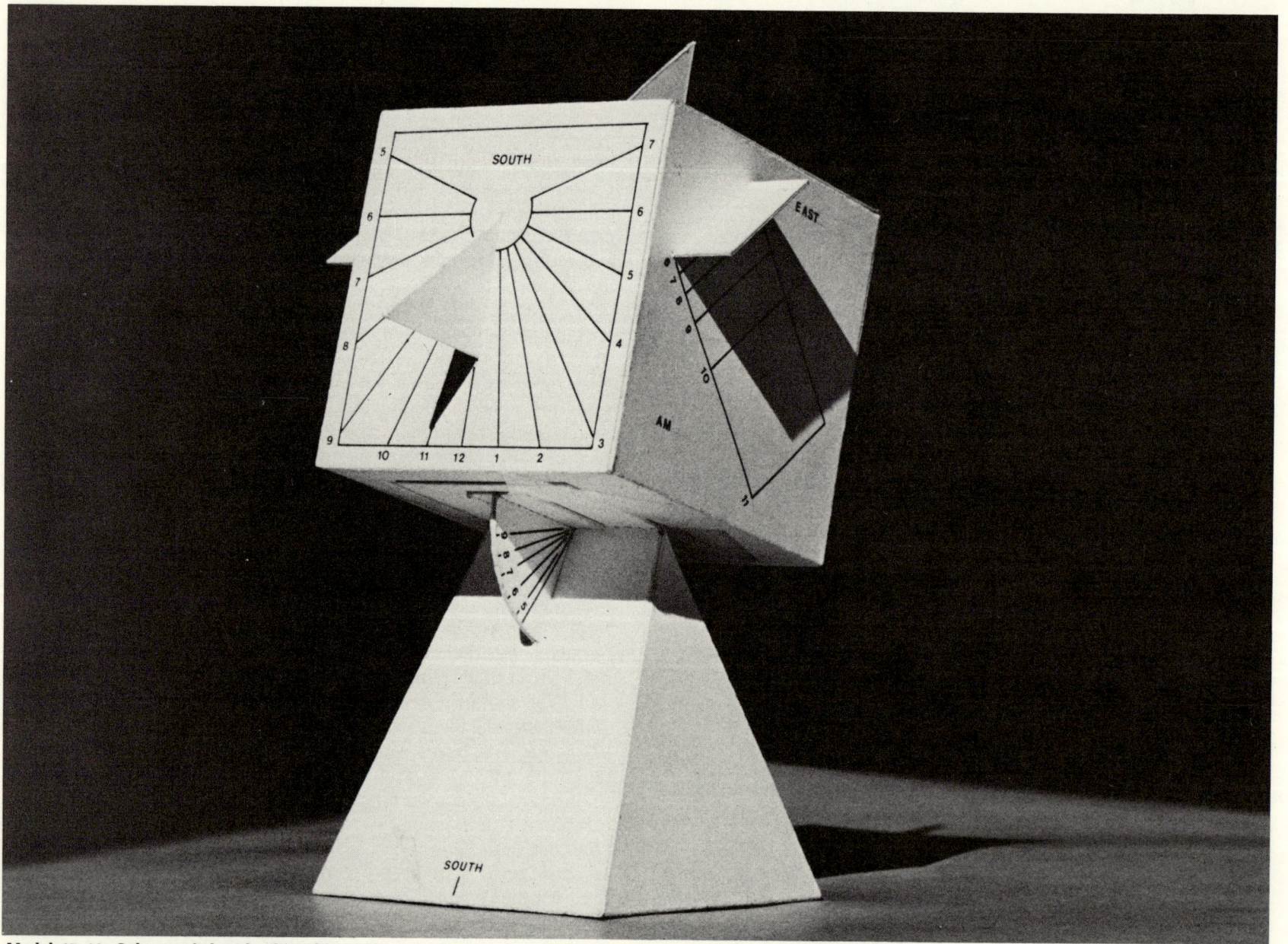

Model 15–16. Cube sundial with tilting base. *Photo by Rosmarie Hausherr.*

Instead of giving you a wedge, as on the horizontal dial, we've provided a tilting stand or base so that your cube will remain truly portable. This is an example of one of many possible accessories that can be invented for the basic dial structure.

The Cube Dial (Putting the Dials Together)

Cut out and assemble plates 15 and 16.

GENERAL INSTRUCTIONS

1. Remove plate 15 from the back of the book before cutting out individual parts.
2. Cut only on gray lines, except when indicated differently.
3. Score all dashed lines (___) on the printed, or top, side of the page. Be careful not to cut through the paper. Score lines with either a pattern-tracing wheel (also called a ponce wheel) or with a dull, slightly rounded tool, such as a butter knife (without teeth).
4. Fold all scored lines back away from the side of the paper containing the score.
5. Use a straightedge as a guide for cutting (with an X-Acto blade or mat knife), scoring, and folding all straight lines for the most precise results.

ASSEMBLY

Cube (A)

1. Cut out A along gray lines.
2. Score all dashed lines (___) according to general instructions.
3. Cut slots in five faces (none in bottom) to the width of two thicknesses of paper for insertion of gnomons.
4. Fold all scored lines, shaping the sides into a cube. Do *not* glue. Set aside.

Gnomons for East and West Sides (B)

1. Cut around B *outside* of the gray lines.
2. Score and fold B in half along line b–b' according to general instructions.
3. *Trim* piece B precisely *on* the gray lines with the halves together so that both sides are identical.
4. Glue the halves of B together and let it dry flat.
5. With the cube resting on its bottom side, fold up "EAST" and

"WEST" sides, inserting up to its projecting shoulders the ends of B into the slot of each side, as shown in illustration 1.

6. Glue B on the inside of the cube and let dry. (A toothpick is a good tool for applying glue to the slot seam.)

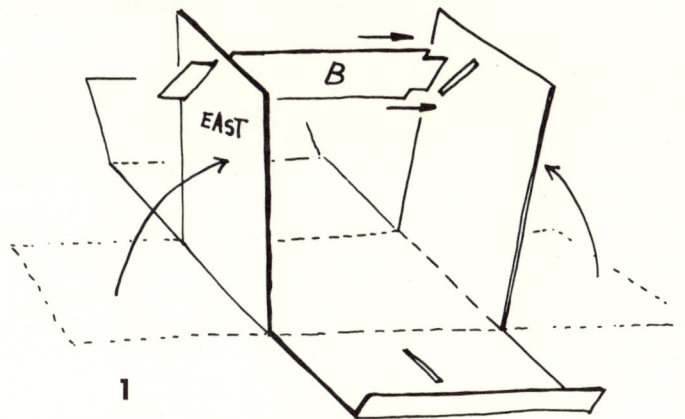

Gnomons for North and South Sides (C)

1. Cut around C *outside* of the gray lines.
2. Score and fold C in half along line c–c' according to general instructions.
3. *Trim* piece C precisely *on* the gray lines with the halves together so that both sides are identical. Cut slot in C.
4. Glue the halves of C together and let it dry flat.
5. With the cube resting on its bottom side, hook B inside of the cube onto the slot of C. Fold up "NORTH" and "SOUTH" sides so that the ends of C fit through the slots, as shown in illustration 2.

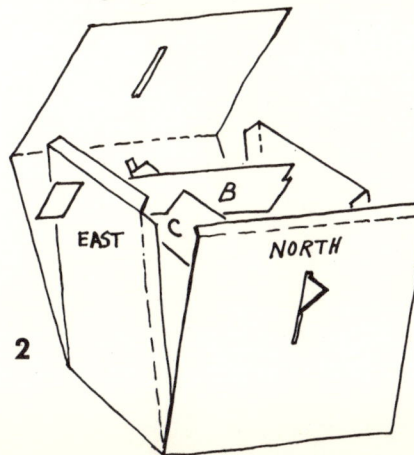

6. Place glue on tabs A, A', B, and B' on the "WEST" and "EAST" sides of the cube. Close up the "NORTH" and "SOUTH" sides (making sure the cube sides rest snugly against the projecting shoulders of the gnomons: Glue on shoulders of C and intersection of B and C is helpful). Hold the sides of the cube closed until the glue has dried.

Gnomon for Top of Cube (D)

1. Score D on dashed lines *before* cutting it out of the page (it is easier to handle when not first detached because of its small size).
2. Cut out D in two pieces along line d–d'.
3. Fold scored lines and then glue the two halves together, as shown in illustration 3.

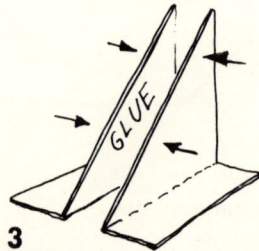

3

4. Insert D through slot on top of the cube (with the apex of the triangle toward the "NORTH" side) and glue it to the blank undersurface there, as shown in illustration 4. Let it dry.
5. Place glue on tabs C^1, C^2, and C^3. Close lid down onto these tabs and hold until glue dries.

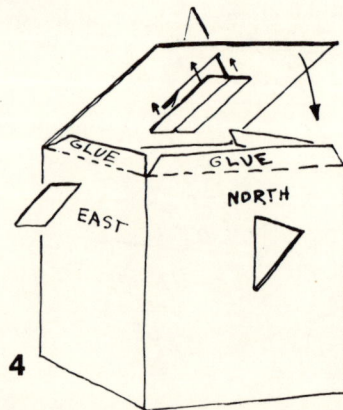

4

HOW TO USE YOUR CUBE DIAL

1. All the dials on this cube have been cut for latitude 45° and each face that gets sunlight should read the same time.
2. If you are at latitude 45°, position the cube so that the twelve o'clock line on the "SOUTH" side lines up with true south and the center of the "NORTH" side lines up with true north. Read the hours at the edges of the shadows. You do not *need* to make the base, although the cube looks best on it.
3. If your latitude is more or less than 45°, you will have to make the tilting base in plate 16 to adjust the angle of your cube to your location.

TILTING BASE FOR CUBE SUNDIAL

GENERAL INSTRUCTIONS

1. Remove plate 16 from the back of the book before cutting out parts.
2. Cut only on gray lines.
3. Score all dashed lines (_ _ _) on the printed, or top, side of the page. Score all dotted lines (. . .) on the blank side, or underside, of the page. Be careful not to cut through the paper. Score lines with either a pattern-tracing wheel (also called a ponce wheel) or with a dull, slightly rounded tool, such as a butter knife (without teeth).
4. Fold all scored lines back away from the side of the paper containing the score.
5. Use a straightedge as a guide for cutting (with an X-Acto blade or mat knife), scoring, and folding all straight lines for the most precise results.

ASSEMBLY

Latitude Wheel (A)

1. Cut out A (including the center slot) before scoring it.
2. On the back, or unprinted, side of the piece, score dotted lines.
3. Cut A in half along line a–a'.
4. Fold the scored line of each piece.
5. Glue both halves together, leaving the folded tabs A and A' free of glue, as shown in illustration 1. Set A aside.

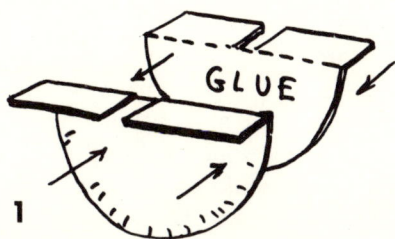

Base (B)

1. Cut out B (including the slots). Use an extra sharp blade on the slots. These slots should be just wide enough to receive the assembled piece A, and the fit should be snug.
2. Score all lines as indicated in the general instructions.

Assembly of A and B Together

1. Fold up the "NORTH" and "SOUTH" sides of B (including the big tabs B and B' so that they meet and their slots align, as shown in illustration 2.

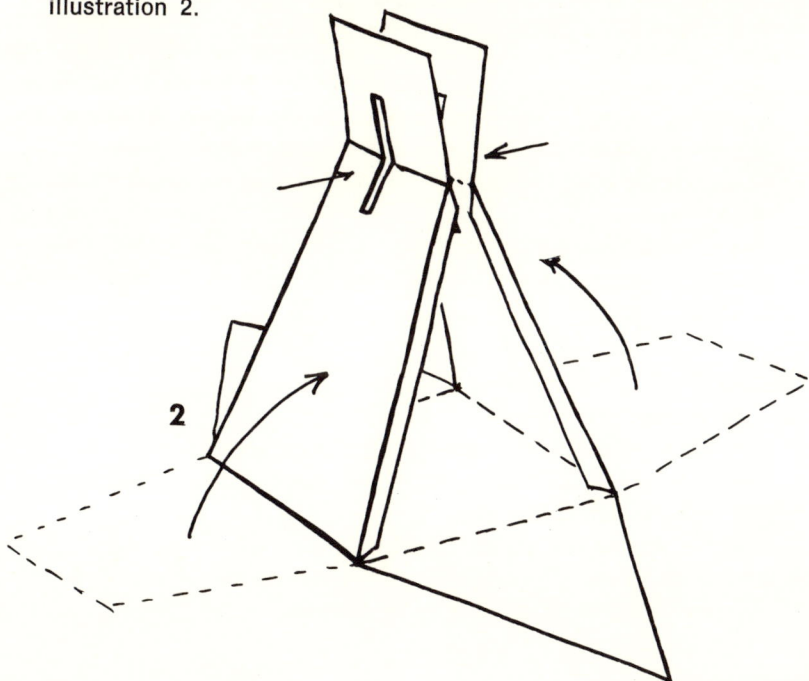

2. Insert A with its tabs folded up halfway into the (common) slots of the "NORTH" and "SOUTH" sides of B, so that the 45° to 90° marks on A are over the "SOUTH" side of B, as shown in illustration 3.

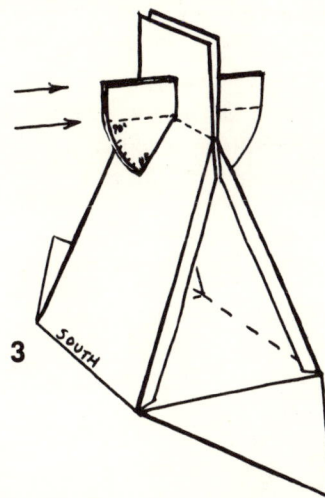

3. Fold down the tabs of A to keep the "NORTH" and "SOUTH" sides of B together.
4. Fold down big tabs on B and glue them to the tabs of A.
5. Place glue on the side tabs of the "NORTH" and "SOUTH" panels of B and fold up the triangular sides to meet them. Hold sides together until glue dries. See illustration 4.

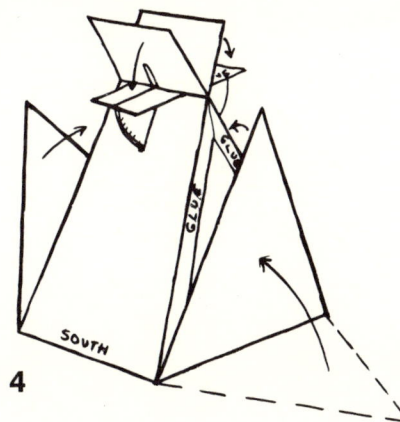

93

HOW TO USE TILTING BASE WITH YOUR CUBE DIAL

1. If your latitude number is *not* 45°, glue or tape your cube to the top of the base so that the "NORTH" and "SOUTH" sides of the cube correspond to the "NORTH" and "SOUTH" sides of the base.

2. Tilt the cube on the base to your latitude number, lining up your latitude line on the circular disc with the edge of the triangular side of the base. Illustration 5 shows the position for latitude 50°.

3. Position the entire cube with base so that the "NORTH" and "SOUTH" marks on your base line up with the true north–south meridian line.

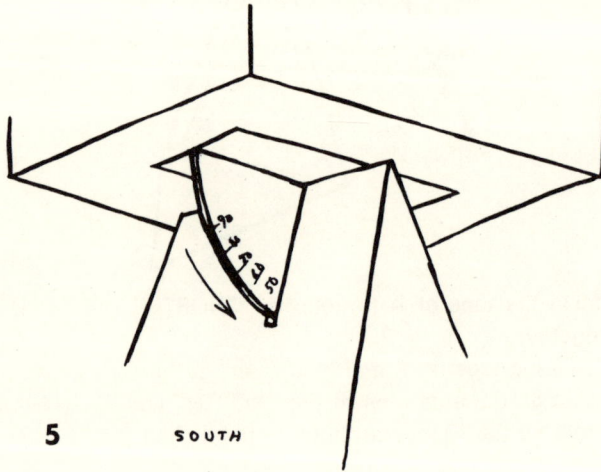

5   SOUTH

With only a few variations of simple basic dial forms you can multiply the faces and shapes of the objects upon which they are mounted (as in our cube), thereby creating many other possible combination dials. We have only started to introduce you to the study of gnomonics, or sundial making, and hope that you will continue, on your own, your exploration of both the sundial and your own living relationship to the sun itself.

Naturally, as these dials are of paper and in miniature they will not offer absolute accuracy. The smaller the dial, the more difficult precision becomes. Therefore, don't expect an astronomical tool here, only approximate hours. The Chinese, Hindu Indians, and Arabs built gnomons up to forty feet tall in their observatories to obtain greater precision, and one early Chinese emperor actually regulated the standard ten- to twelve-foot height of all imperial dial gnomons. Our dials, on the other hand, aim to capture the spirit and poetry of cyclic time and the beauty of sculptured geometry in sunlight rather than—how shall we put it?—the dial's grammar and correct punctuation. Our own attitude towards sundials is an expansive one. We enjoy their present time-telling aspect but we see it as secondary to their ability to inspire inquiry and contemplation of life beyond the clock.

WITH THE SUNLIGHT
AND FOR THIS SPECIAL PLACE
AS A BRIDGE
BETWEEN HEAVEN AND EARTH
AT THIS COORDINATE IN TIME
WE BUILD

Robert Adzema
Mablen Jones

Bibliography

Binder, Pearl. *Magic Symbols of the World.* London–New York, 1972.

Blackburn, Thomas C., ed. *December's Child—A Book of Chumash Oral Narratives.* Berkeley–Los Angeles–London, 1975.

Branley, Franklyn. *The Sun—Star Number One.* New York, 1964.

Bulfinch, Thomas. *Bulfinch's Mythology.* New York (n.d.).

Cousins, Frank. *Sundials—The Art and Science of Gnomonics.* New York, 1970.

Curtin, Jeremiah. *Creation Myths of Primitive America.* New York–London, 1898 (reissued, Bronx, N.Y., 1969).

Dantzig, Tobias. *Number—The Language of Science.* New York, 1954.

de Santillana, Giorgio, and von Dechend, Hertha. *Hamlet's Mill—An Essay on Myth and the Frame of Time.* Boston, 1969.

Dolan, Winthrop W. *A Choice of Sundials.* Brattleboro, Vt., 1975.

Earle, Alice Morse. *Sundials and Roses of Yesterday.* New York, 1902 (reissued, New York, 1969).

Encyclopaedia Britannica, vols. 2 and 21. Chicago–London, 1973.

Engelbrektson, Sune. *Stars, Planets, and Galaxies.* New York–Toronto–London, 1975.

Farrington, Benjamin. *Science in Antiquity.* New York–London–Oxford, 1969.

Franklin, Kenneth L. "The Astronomer's Odd Figure 8." *Natural History* 71 (Oct. 1962): 8–15.

Fraser, J. T., ed. *The Voices of Time—A Cooperative Survey of Man's Views of Time as Expressed by the Sciences and by the Humanities.* New York, 1966.

Frazer, Sir James George. *The Golden Bough—A Study in Magic and Religion.* New York, 1937.

Gauquelin, Michel. *Cosmic Clocks—From Astrology to a Modern Science.* Chicago, 1967.

Gayley, Charles Mills. *The Classic Myths in English Literature and in Art.* New York–London, 1939.

Govinda, Lama Anagarika. *Psycho-Cosmic Symbolism of the Buddhist Stupa.* Emeryville, Calif., 1976.

Guye, Samuel, and Michel, Henri. *Time and Space—Measuring Instruments from the Fifteenth to the Nineteenth Century.* New York, 1971.

Hawkins, Gerald, and White, John B. *Stonehenge Decoded.* New York, 1965.

Helfman, Elizabeth S. *Signs and Symbols Around the World.* New York, 1967.

Kline, Morris. *Mathematics and the Physical World.* New York, 1959.

Kyselka, Will, and Lanterman, Ray. *North Star to Southern Cross.* Honolulu, Hawaii, 1976.

Lee, Sherman E. *A History of Far Eastern Art.* Englewood Cliffs, N.J.–New York (n.d.).

Liu, Da. *I Ching Coin Prediction.* New York, 1975.

Luce, Gay Gaer. *Body Time—Physiological Rhythms and Social Stress.* New York, 1971.

McKeon, Richard, ed. *Introduction to Aristotle.* New York, 1947.

Marshall, Roy K. *Sundials.* New York, 1963.

Mayall, R. Newton, and Mayall, Margaret. *Sundials—How to Know, Use, and Make Them.* Boston, 1962.

Merleau-Ponty, Jacques, and Morando, Bruno. *The Rebirth of Cosmogony.* New York, 1976.

Moore, Patrick. *The Sun.* New York, 1968.

Needham, Joseph. *Mathematics and the Sciences of the Heavens and the Earth.* Science and Civilisation in China, vol. 3. New York–London: 1959.

Odishaw, Hugh, ed. *The Earth in Space.* New York–London, 1967.

Olcott, William Tyler. *Myths of the Sun—A Collection of Myths and Legends Concerning the Sun and Its Worship.* New York, 1967. (Original title: *Sun Lore of All Ages.* New York, 1914.)

Oliver, Bernard M. "The Shape of the Analemma." *Sky and Telescope* 44 (July 1972): 20–22.

Perry, W. J. *The Children of the Sun—A Study in the Early History of Civilization.* Grosse Point, Mich., 1968 (originally London, 1923).

Price, Derek de Solla. *Gears From the Greeks—The Antikythera Mechanism, A Calendar Computer from ca. 80 B.C.* New York, 1975.

Rohr, Rene R. J. *Sundials—History, Theory, and Practice.* Toronto, 1970. (Original title: *Les Cadrens Solaires.* 1965.)

Schuyler, Garret L. "Amateur Scientist (How to Make a Ceiling Sundial)." *Scientific American 194* (March 1956): 148–50.

Smith, Richard Furnald. *Prelude to Science—An Exploration of Magic and Divination.* New York, 1975.

Spence, Lewis. *Myth and Ritual in Dance, Game, and Rhyme.* London, 1947 (reissued, Detroit, 1971).

Waugh, Albert E. *Sundials—Their Theory and Construction.* New York, 1973.

Whitrow, G. J. *The Nature of Time.* New York–Chicago–San Francisco, 1973. (Original title: *What Is Time?* London, 1972.)

Wright, Lawrence. *Clockwork Man—The Story of Time, Its Origins, Its Uses, Its Tyranny.* New York, 1968.

Young, Louise B., ed. *Exploring the Universe.* New York, 1963.

INDEX

SOUTH

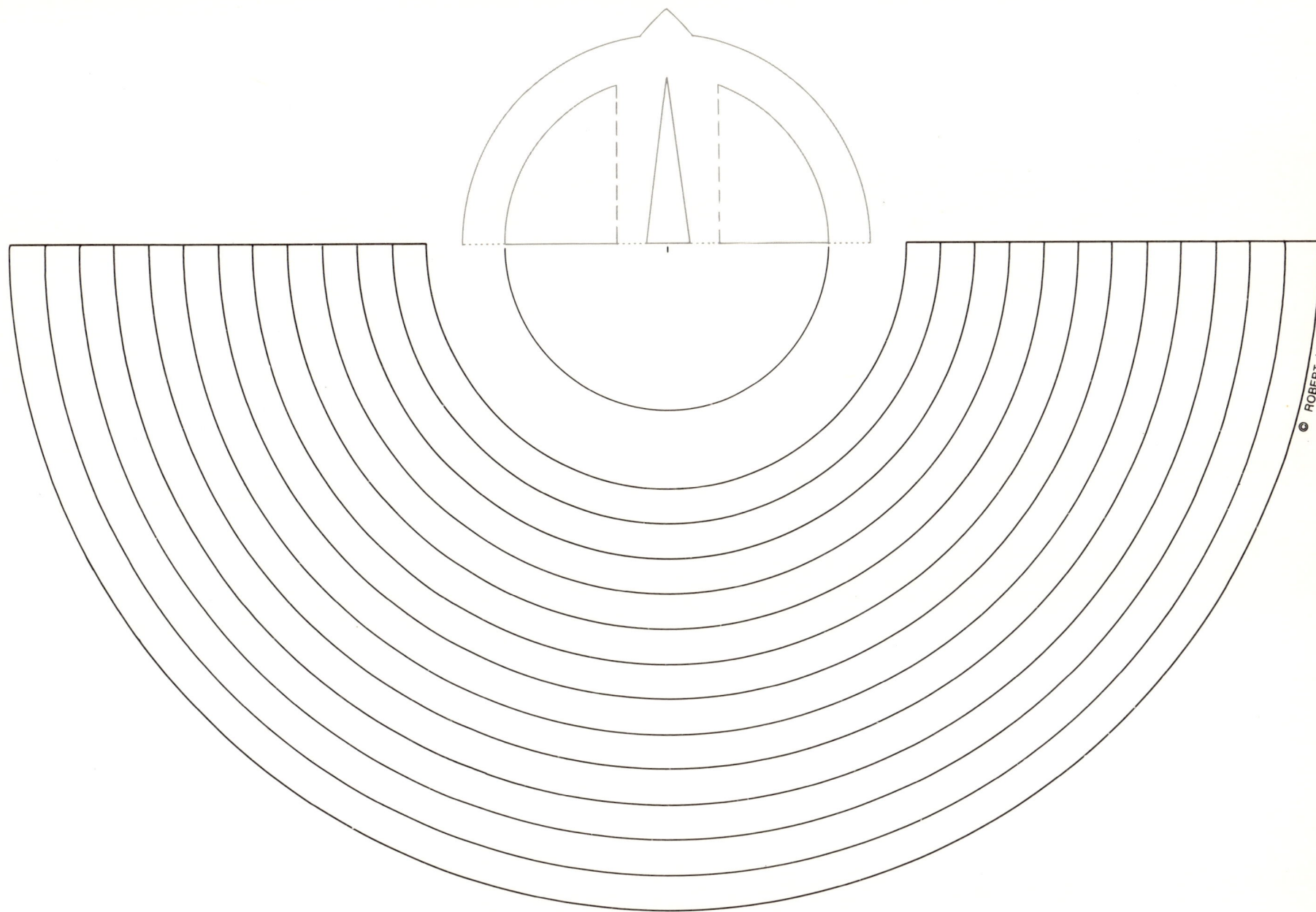

ROBERT ADZEMA 1978

PLATE 2

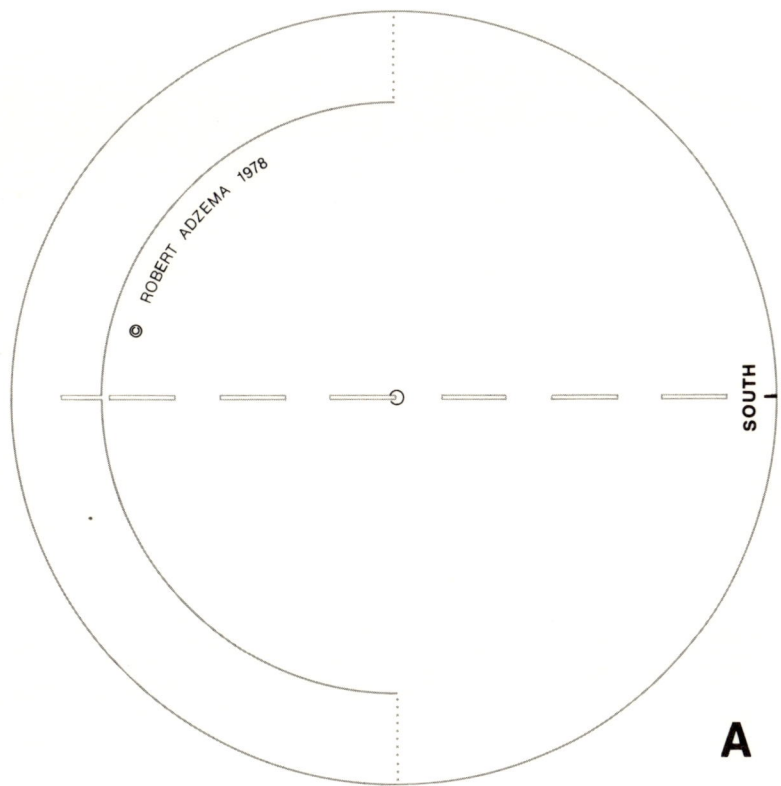

A

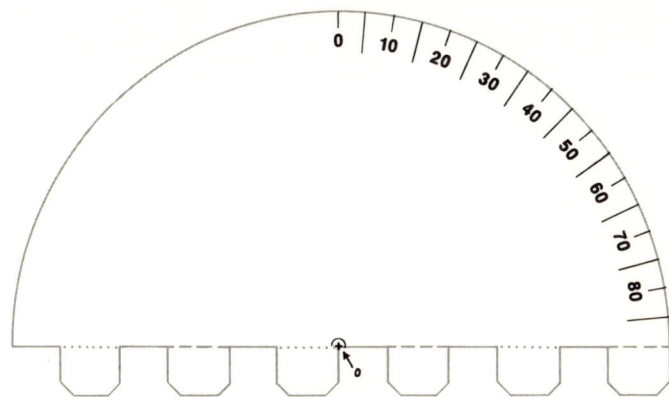

B

PLATE 3

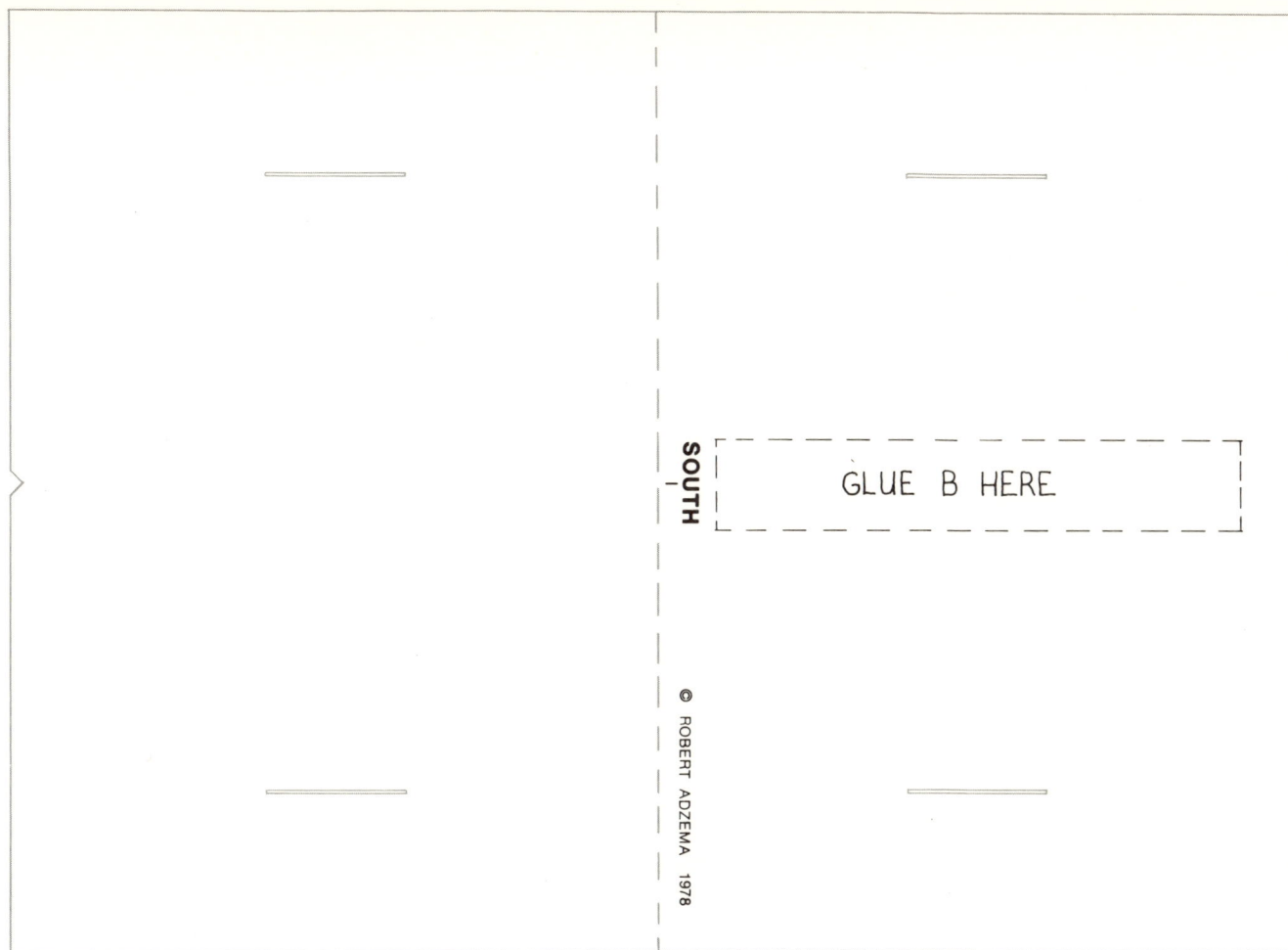

PLATE 4

SOUTH

GLUE B HERE

© ROBERT ADZEMA 1978

A

A

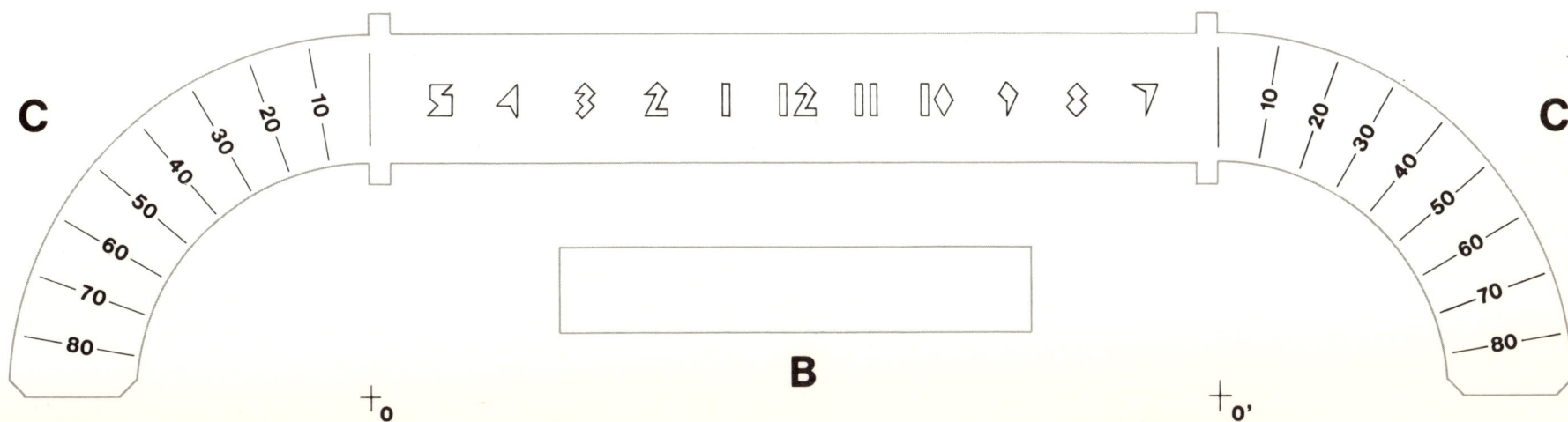

C

C

B

+₀   +₀'

TAB A

SIDE A

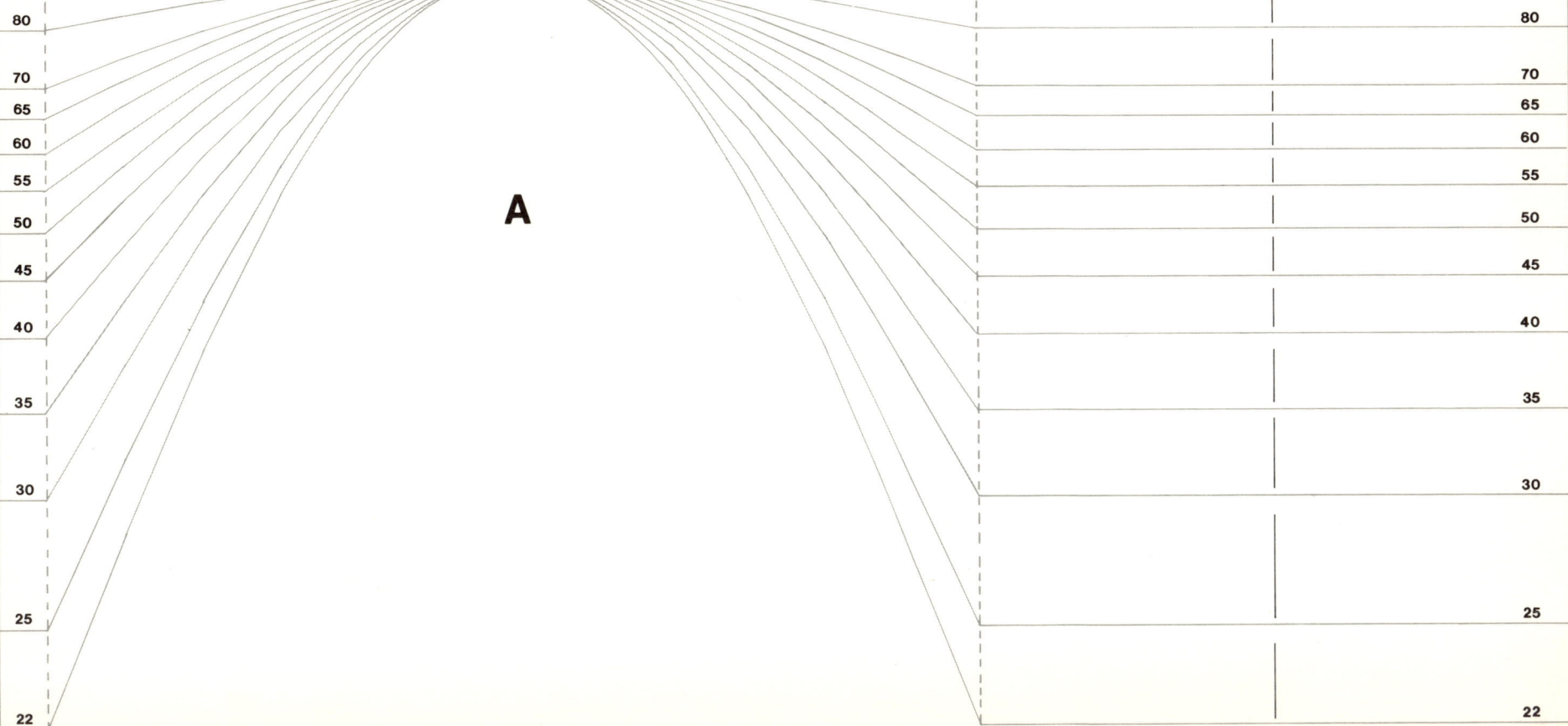

REMOVE

SOUTH
LINE

N

A

80	80
70	70
65	65
60	60
55	55
50	50
45	45
40	40
35	35
30	30
25	25
22	22

PLATE 5

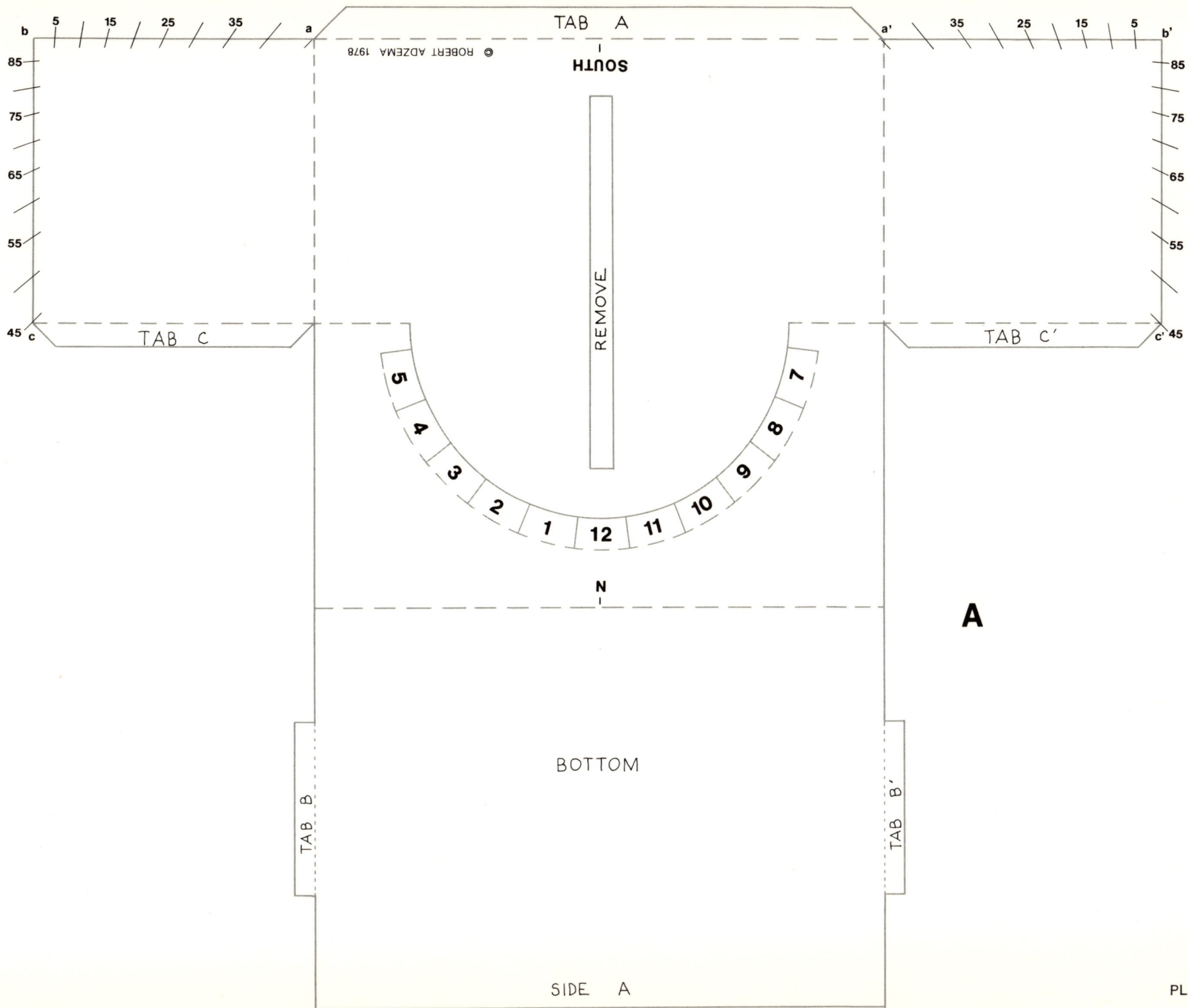

b 5   15   25   35   a   TAB A   a'   35   25   15   5   b'

85   85

75   © ROBERT ADZEMA 1978   75

65   65

                              SOUTH

55   55
                              REMOVE

45 c   TAB C   TAB C'   c' 45

                    5   7
                4   8
                3   9
                2   10
                1   12   11

                                N

                                                                            A

                            BOTTOM

        TAB B   TAB B'

                            SIDE A   PLATE 6

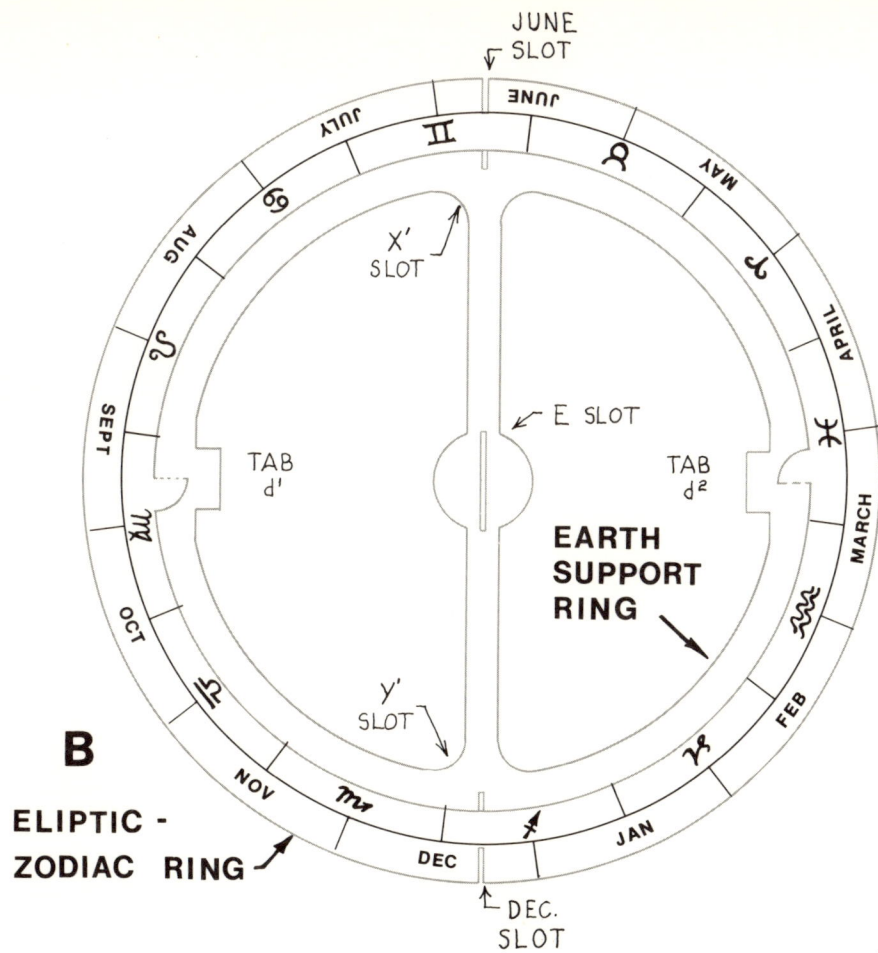

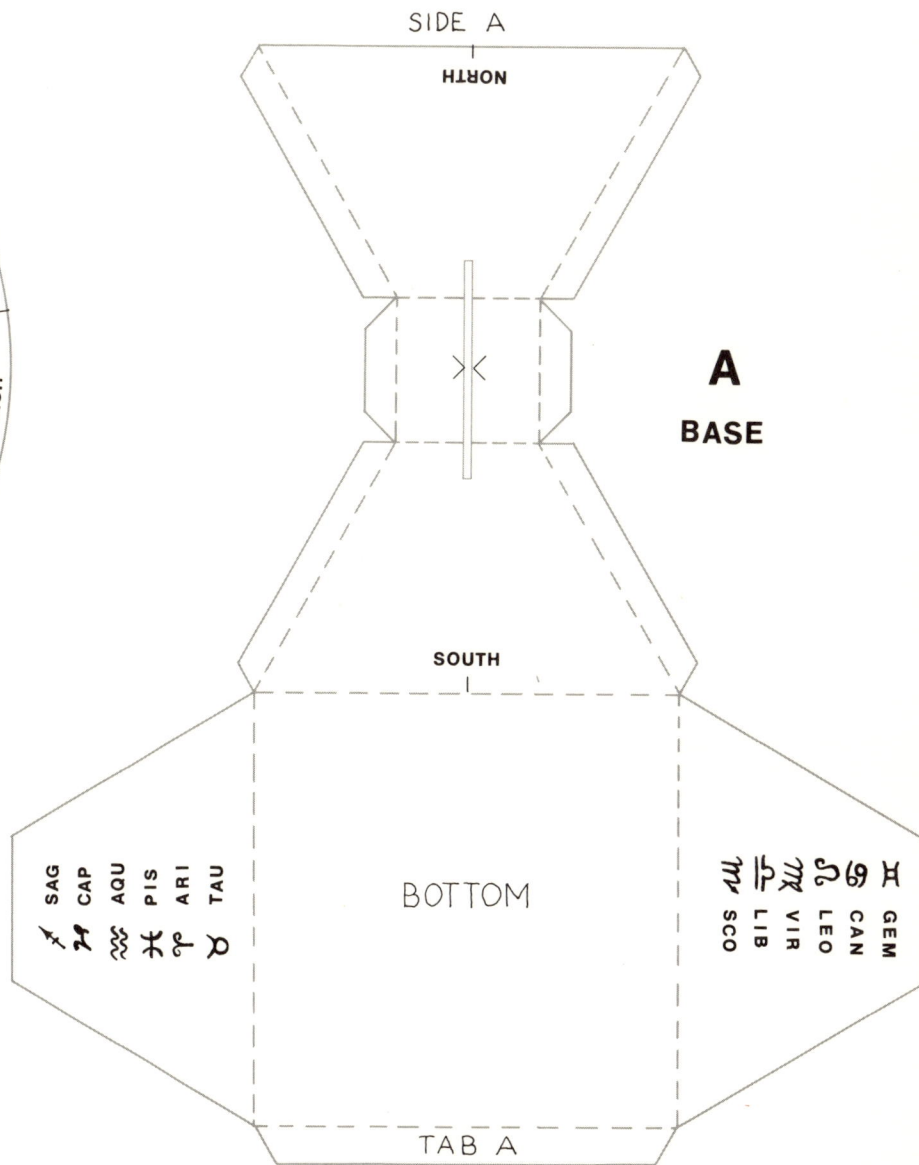

PLATE 7

F
LATITUDE SUPPORT

0
10
20
30
40
50
60
70
80
90

y
SLOT

ECLIPTIC POLE

NORTH CELESTIAL POLE

EP

90 NCP

X
SLOT

JUNE
SLOT

TAB
d^3

TAB
d^4

0

10

20

30

40

50

60

70

80

90

DEC
SLOT

y
SLOT

© ROBERT ADZEMA 1978

0

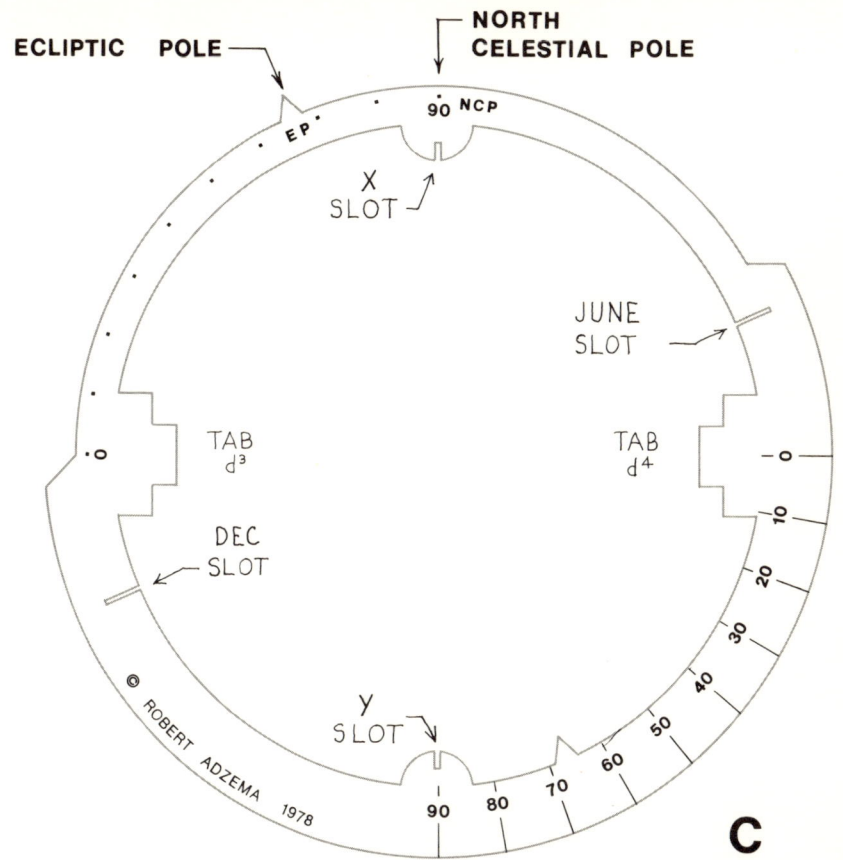

C
LATITUDE RING

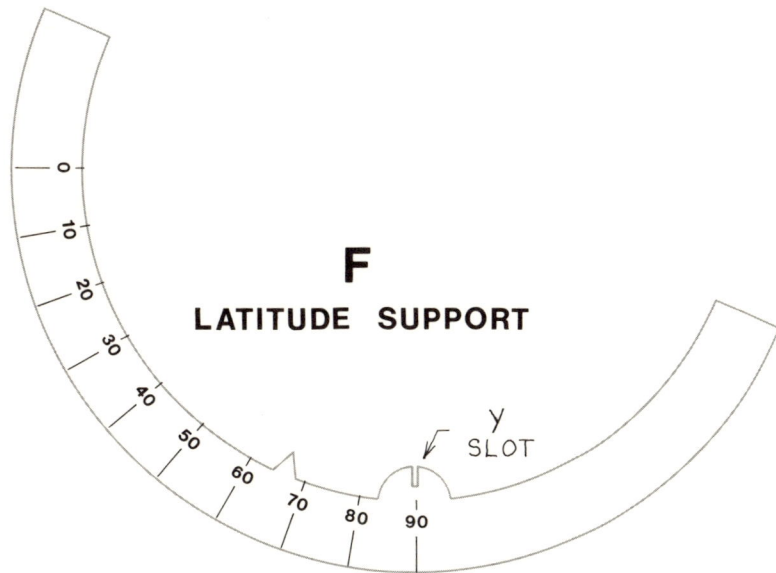

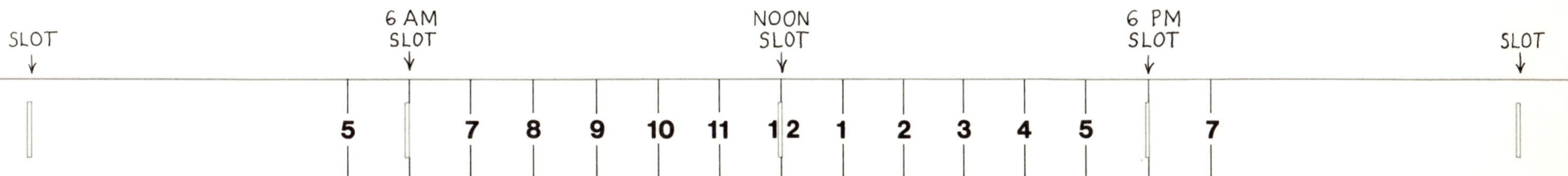

SLOT

6 AM
SLOT

NOON
SLOT

6 PM
SLOT

SLOT

5   7   8   9   10   11   12   1   2   3   4   5   7

D   EQUATORIAL BAND

PLATE 8

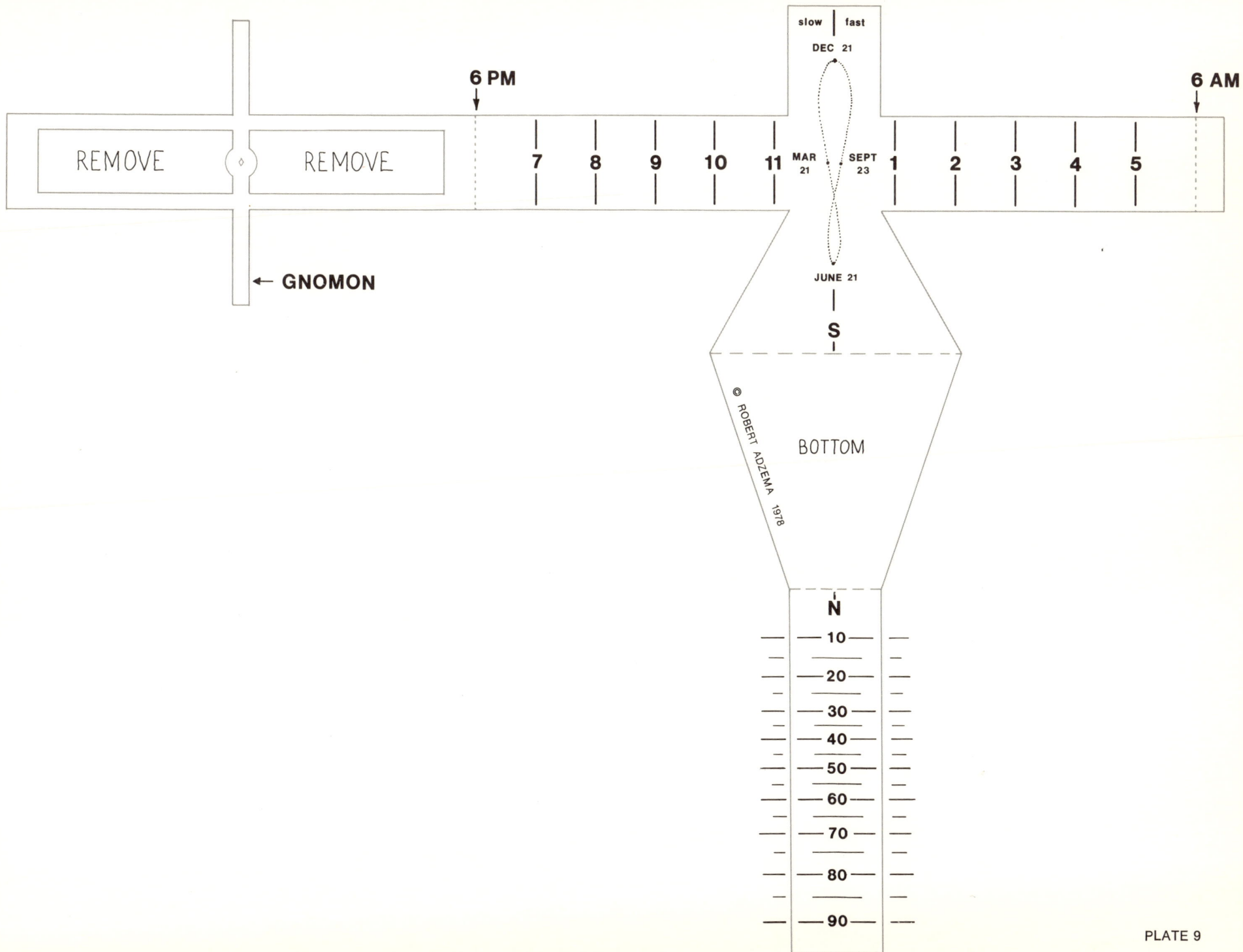

slow | fast

DEC 21

6 PM

6 AM

REMOVE   REMOVE

7   8   9   10   11

MAR 21   SEPT 23

1   2   3   4   5

← GNOMON

JUNE 21

S

© ROBERT ADZEMA 1978

BOTTOM

N

10
20
30
40
50
60
70
80
90

PLATE 9

SOUTH

NORTH

PLATE 10

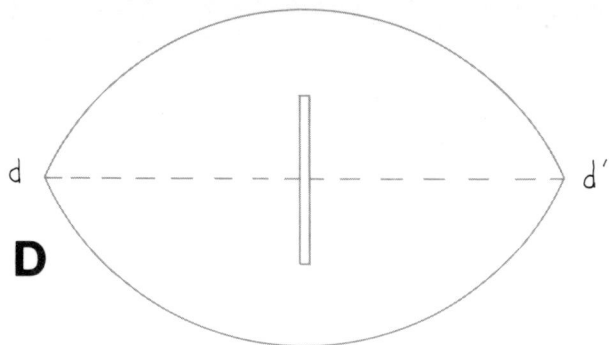

D

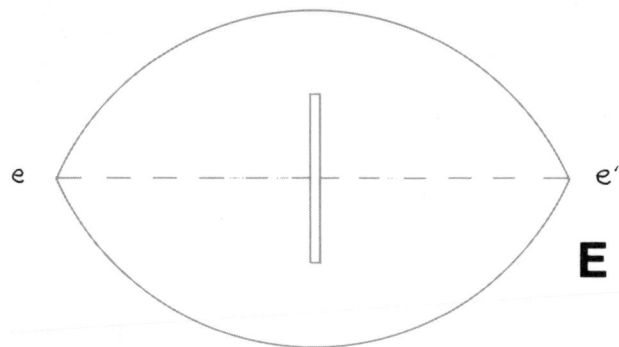

E

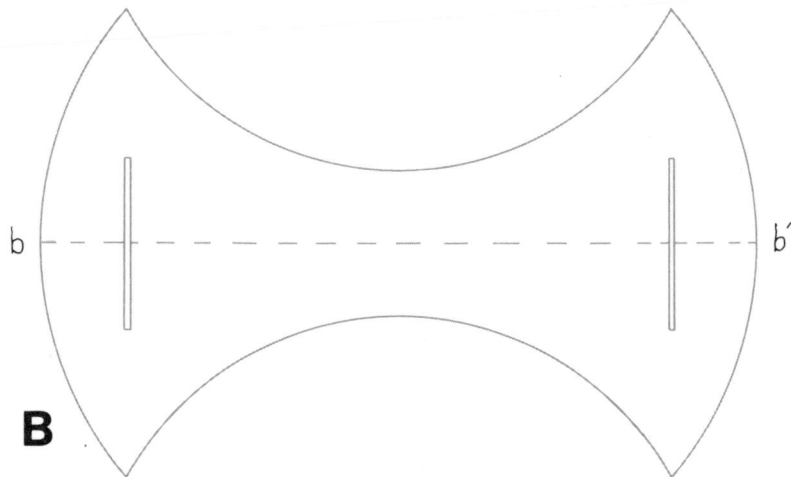

B

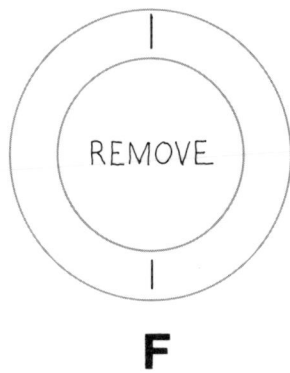

F

REMOVE

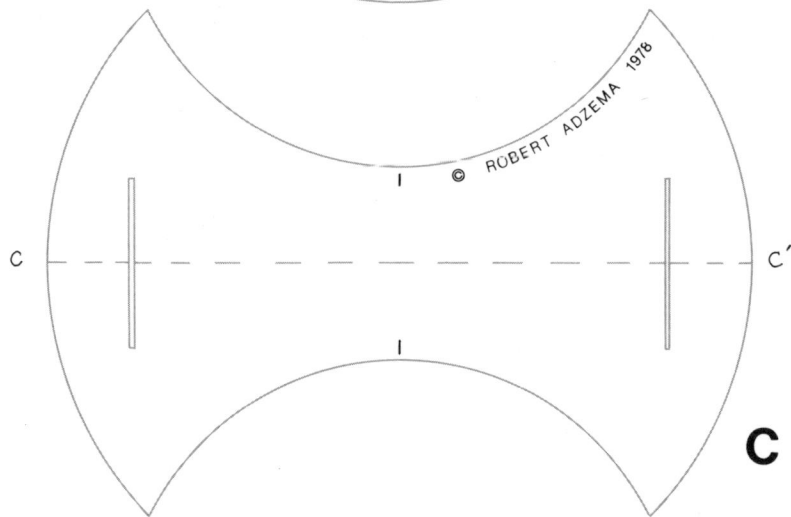

C

© ROBERT ADZEMA 1978

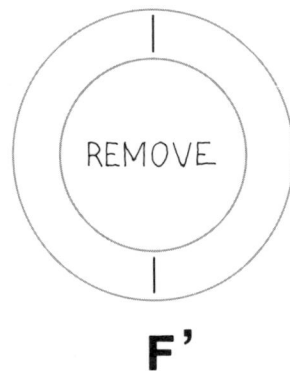

REMOVE

F'

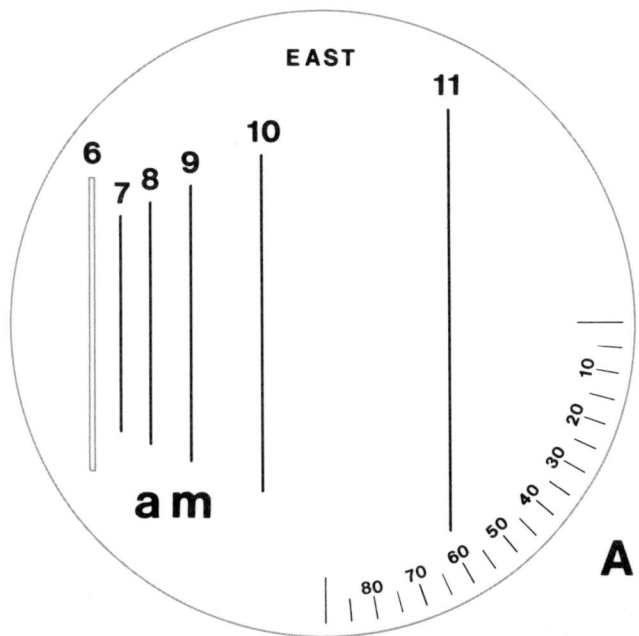

EAST

am

A

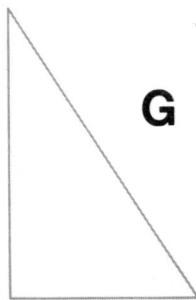

G

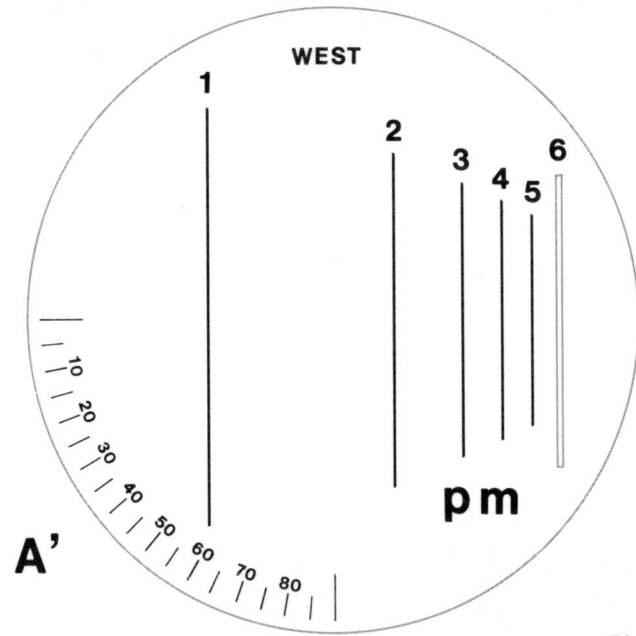

WEST

pm

A'

PLATE 11

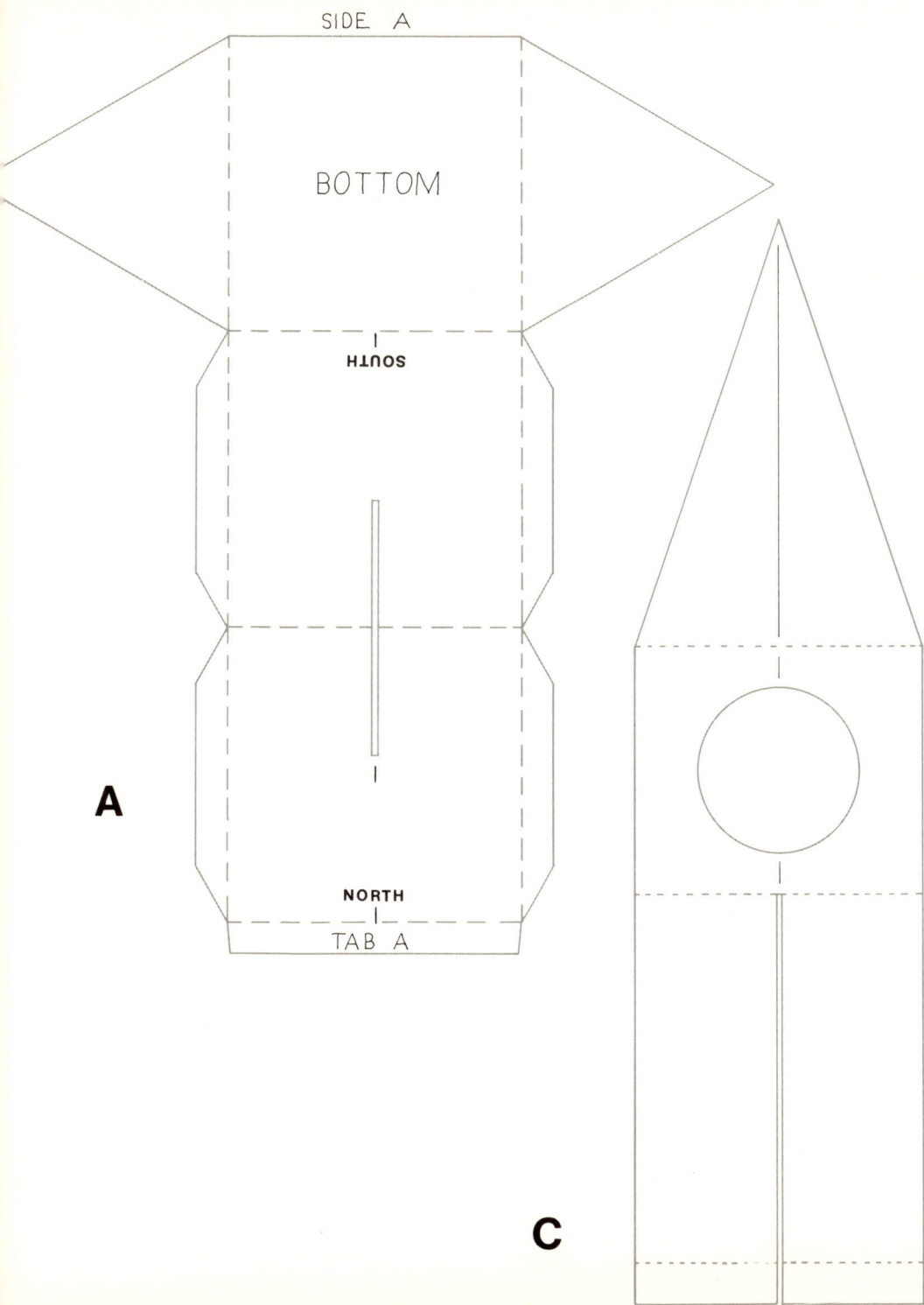

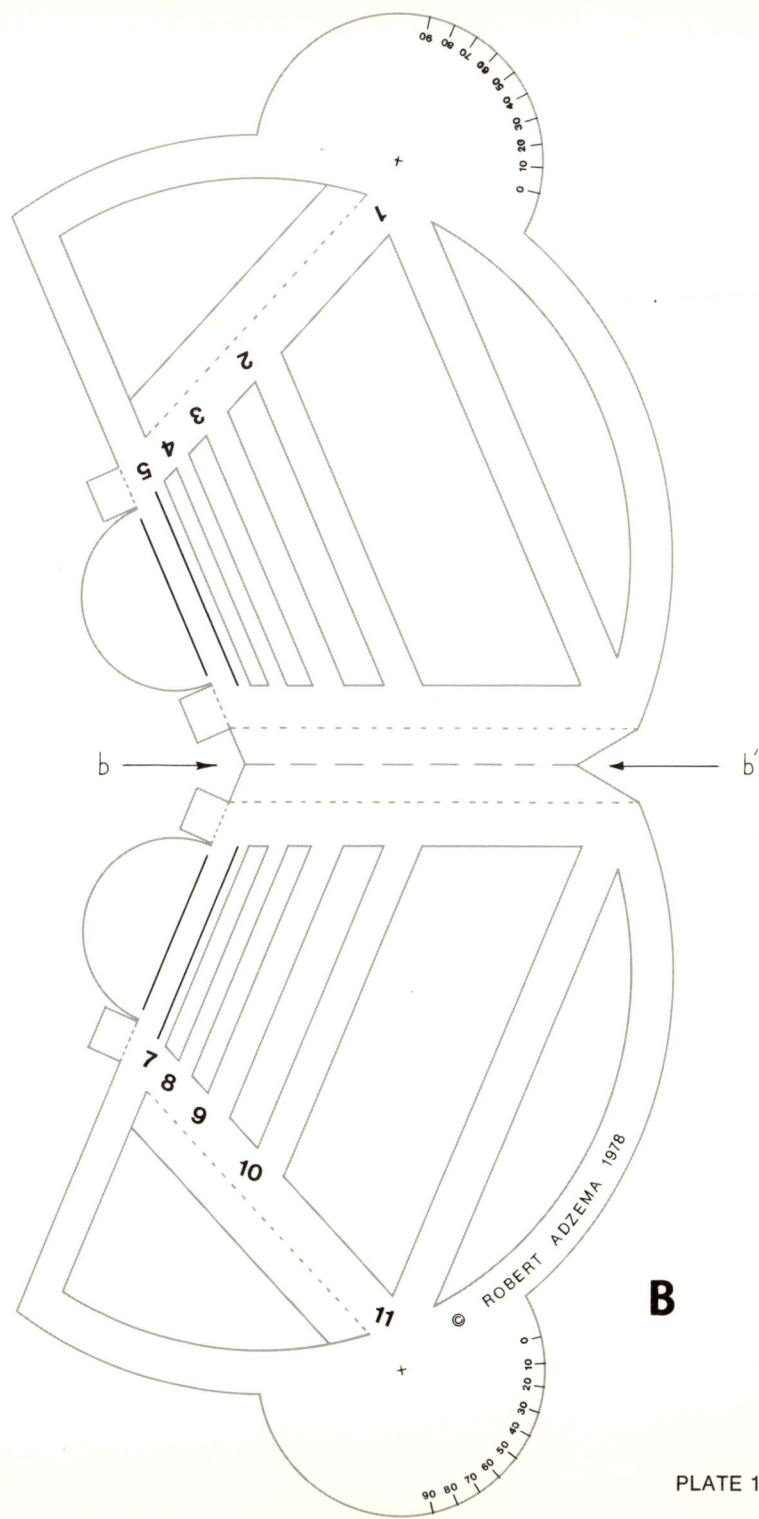

SIDE A

BOTTOM

SOUTH

NORTH

TAB A

A

C

b → ← b'

1
2
3
4
5

7
8
9
10
11

© ROBERT ADZEMA 1978

B

PLATE 12

C

UNIVERSAL WEDGE

GLUE "**B**" HERE
IF YOUR LATITUDE IS
MORE THAN 5° FROM
40°

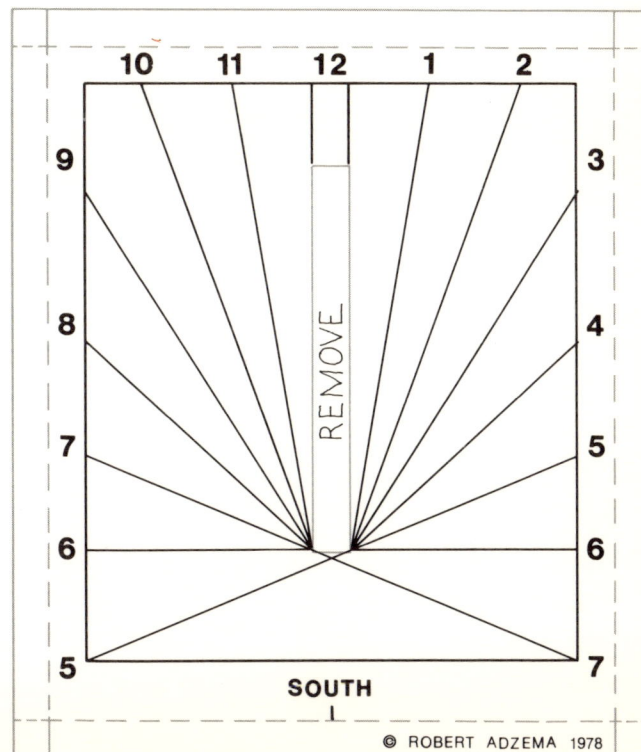

© ROBERT ADZEMA 1978

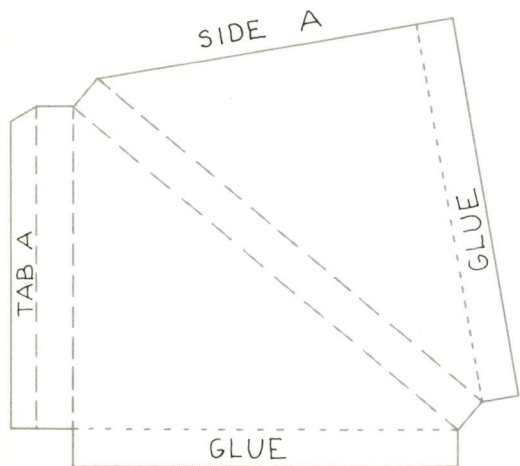

40 35 30 25 20 15 10 5

5
10
15
20
25
30
35
40
45

SIDE A

TAB A

GLUE

GLUE

A **GNOMON**

DIAL TABLE **B** →

10 11 12 1 2

9   3

8   REMOVE   4

7   5

6   6

5   7

SOUTH

© ROBERT ADZEMA 1978

PLATE 13

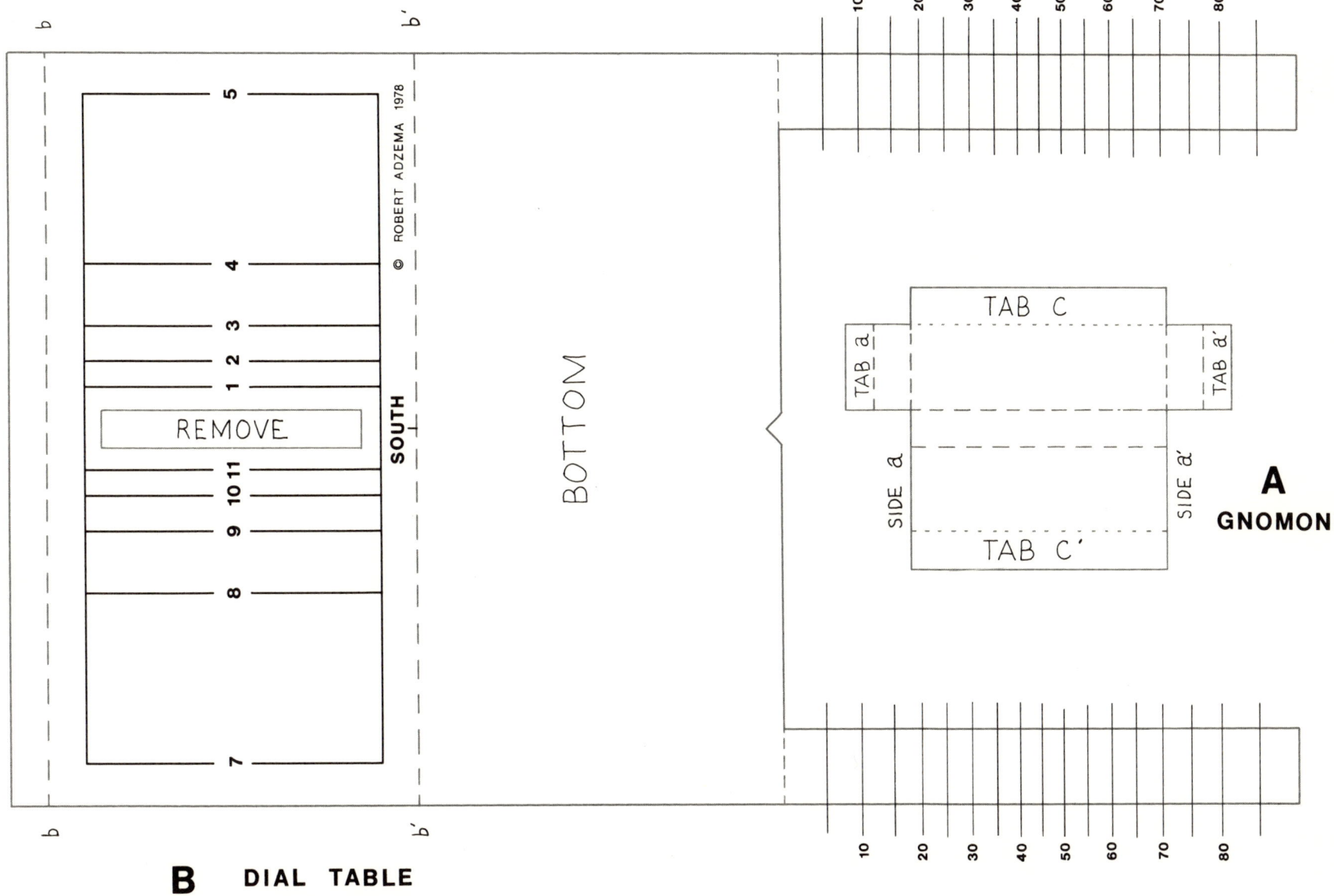

A **GNOMON**

TAB C

TAB a

TAB a'

SIDE a

SIDE a'

TAB C'

© ROBERT ADZEMA 1978

BOTTOM

SOUTH

REMOVE

b

b'

5

4

3

2

1

10 11

9

8

7

10
20
30
40
50
60
70
80

10
20
30
40
50
60
70
80

B **DIAL TABLE**

PLATE 14

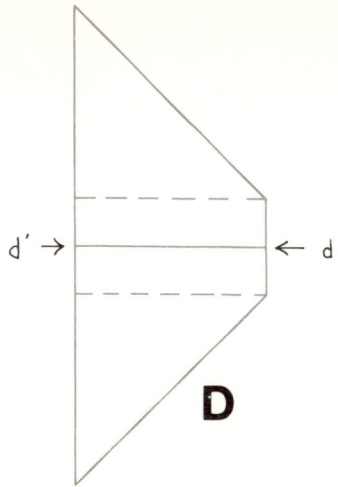

D

d' → ← d

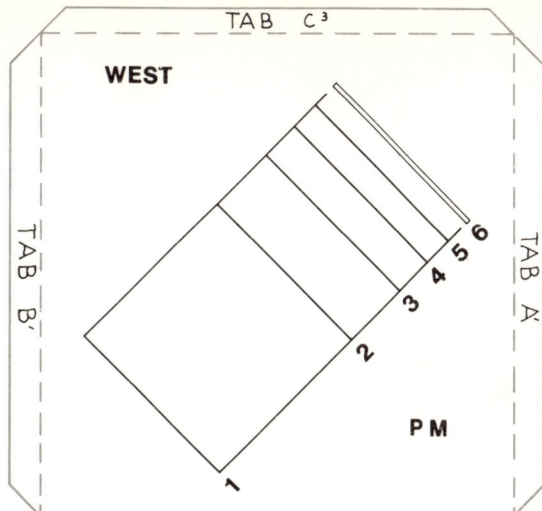

WEST

PM

6 5 4 3 2 1

TAB A'

TAB B'

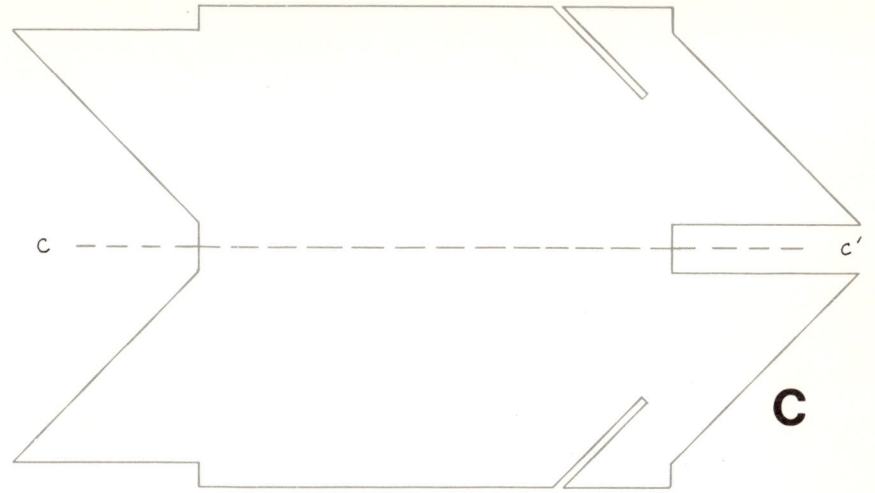

c — — c'

C

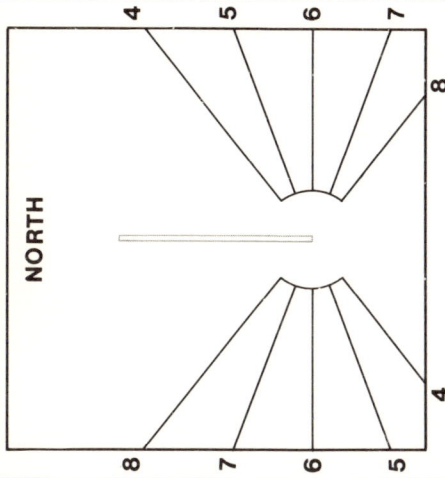

4 5 6 7 8

NORTH

8 7 6 5 4

**CUBE
A**

IF YOUR LATITUDE
IS MORE OR LESS
THAN 45°,
GLUE BASE HERE

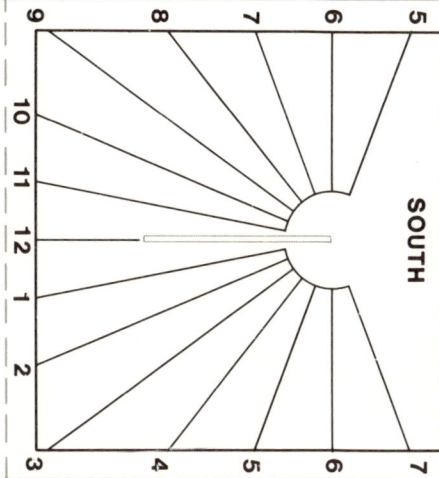

9 8 7 6 5

10

11

12

SOUTH

1

2

3 4 5 6 7

© ROBERT ADZEMA 1978

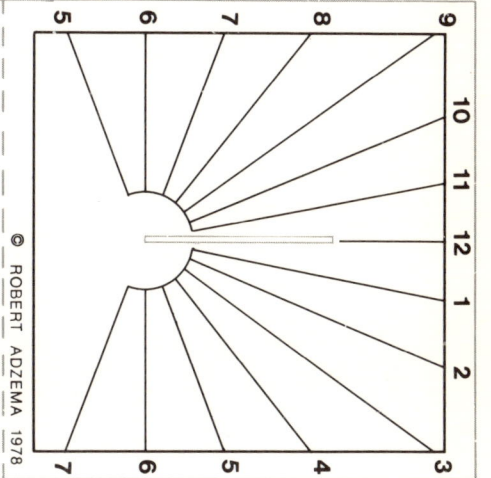

5 6 7 8 9

10

11

12

1

2

7 6 5 4 3

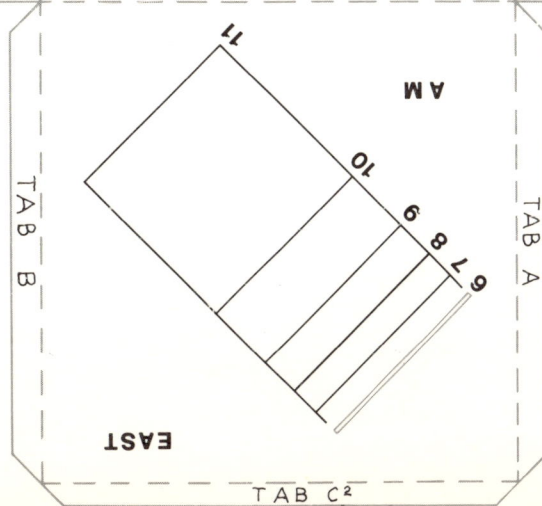

11

10

9 8 7 6

AM

TAB B

TAB A

EAST

TAB C²

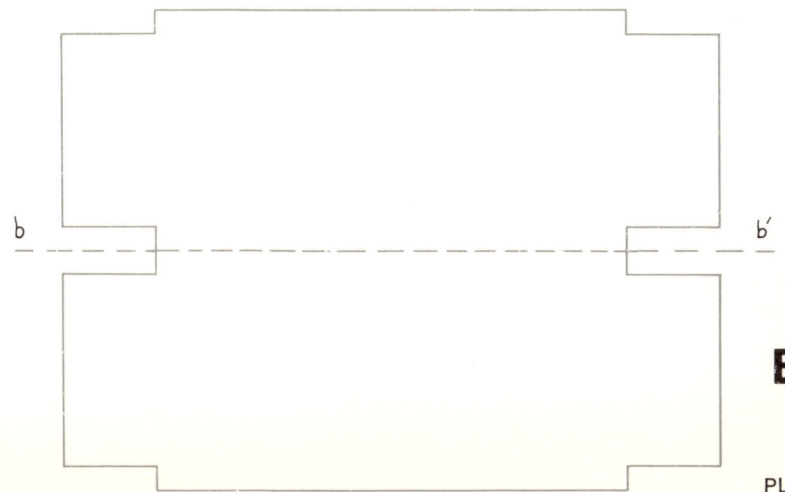

b — — b'

B

PLATE 15

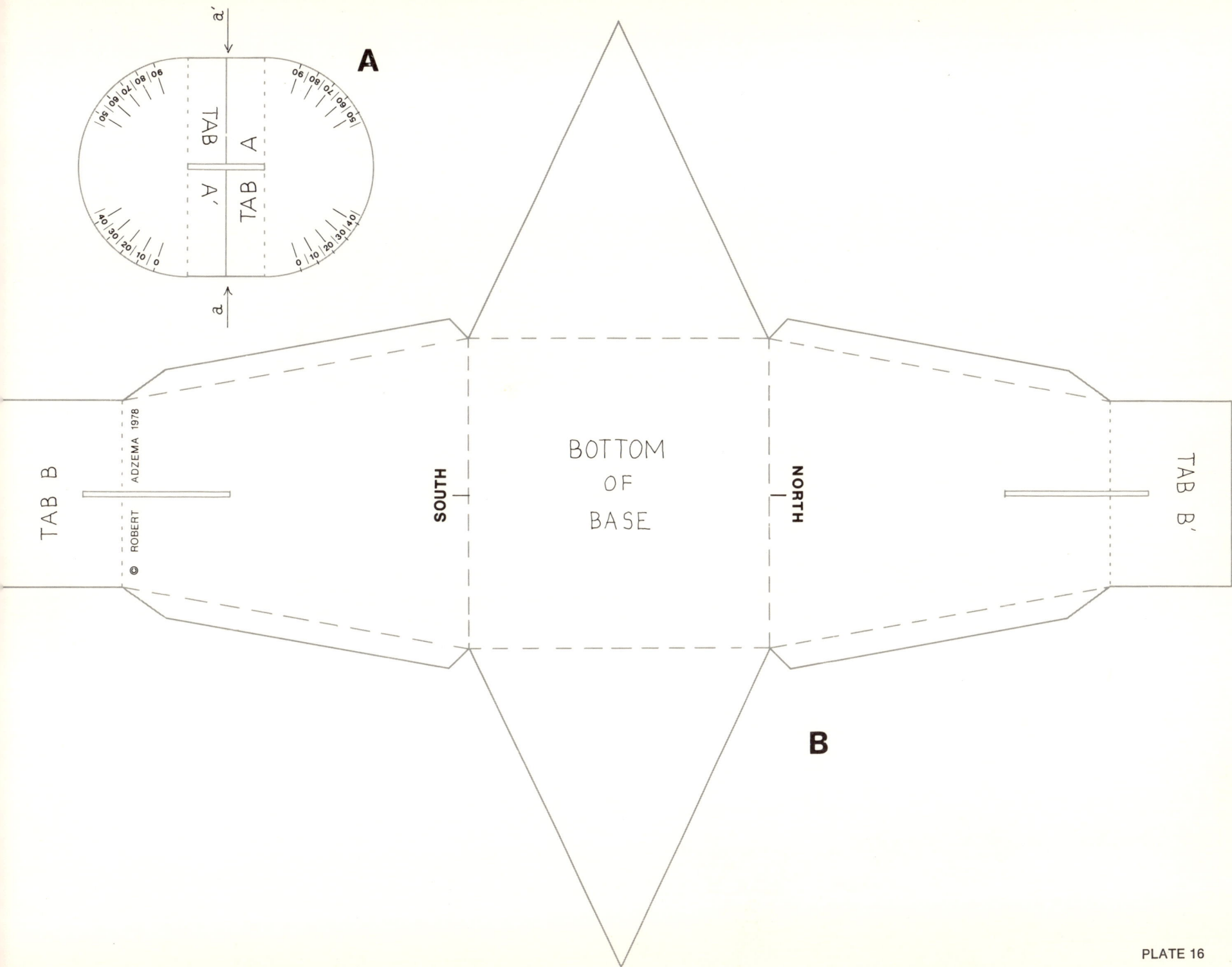

A

a'

TAB A'   TAB A

90 80 70 60 50
50 60 70 80 90
40 30 20 10 0
0 10 20 30 40

a

B

TAB B

© ROBERT   ADZEMA 1978

SOUTH

BOTTOM
OF
BASE

NORTH

TAB B'

PLATE 16